PERPETUAL MOTION

AN ANCIENT MYSTERY SOLVED?

Johann Ernst Elias Bessler, also known as Orffyreus. Above are some examples of his signature. The highly decorative signature was typical of the time, but the rather neat monogram used later on was unique to Orffyreus. It was clearly meant to represent his perpetual motion wheel.

PERPETUAL MOTION

AN ANCIENT MYSTERY SOLVED?

by

John Collins

PERMO PUBLICATIONS

Published and distributed by:
Permo Publications
PO Box 2001, Leamington Spa, CV32 6YQ

ISBN 0 9531615 0 1

British Library Cataloguing in Publication Data.

Dedication

Perpetual Motion:An Ancient Mystery Solved? could not have been written without the everlasting patience, love and warm support of my wife, Sandra. This book is therefore dedicated to her.

Acknowledgements

Perpetual Motion; An Ancient Mystery Solved? represents the culmination of at least ten years research and many years of casual interest. I have received assistance over the years from many people in libraries throughout this country and in Germany and Russia, but I would like to single out for a special thank you, the staff of Rugby library in Warwickshire, those in Birminghan City library and of course the men and women who operate the magnificent facilities at the British Library in London. There are others too numerous to mention, who helped in an amateur capacity and who were able to assist in translating many foreign extracts in both eighteenth century German and Russian.

I would like to express my thanks to the following people for their help and assistance during the writing of this book: Professor Ragnhild Hatton, Dr. H.B. Gardner-McTaggart, Mary Sampson at The Royal Society, Mrs P.M.A.T. Chamberlayne-Macdonald of Cranbury Park, Winchester, Maureen Perrie of Birmingham University, and Ingrid Budden. I must also say a big thank you and acknowledge the help of Michael Senior, BA., without whose sterling work in deciphering and then translating large chunks of eighteenth century German text, much of the research for this book could not have been carried out.

The information about this book which has appeared on the internet would not have been possible without the invaluable work of David Pohl, BEng (Hons), MIEE, PGCE, CNA. He is responsible for the design and authoring of the web pages at http://www.free-energy.co.uk and I owe him a special debt of thanks.

I should not forget to thank my family for tolerating the many hours when I was 'absent', working at the keyboard of my pc, or researching at various archives. They have shown endless patience during this project.

Finally I would like to pay a tribute to all those inventors who were convinced of the possibility of perpetual motion and worked alone and unaided designing and constructing wonderful machines which remained resolutely stationary. Without their untiring efforts there would be no such chimera as a perpetual motion machine.

Design & production: Dick Richardson, Country Books, Little Longstone DE45 1NN

Printed & bound by: MFP Design & Print, Stretford, Manchester M32 0JT

CONTENTS

LIST OF ILLUSTRATIONS

Chapter One

THE LEGEND OF ORFFYREUS' WHEEL

PERPETUAL MOTION

A DEFINITION:
A device which, once having been started, continues to run for ever, without any additional input of energy.

The theory of Perpetual Motion has been consigned to the dustbin of discarded myths; ahead of such pseudo-scientific suppositions as UFO's, crop circles, holistic medicine and all the paraphernalia of the paranormal. The mighty weight of established scientific opinion is against such beliefs, because they cannot be explained within current scientific dogma. The dictionary definition of *dogma* is - 'settled opinion . . . indisputable doctrine'. So theories about certain known facts are accepted and indisputable, and *incapable* of modification. This rigid adherence to current theories becomes unstuck only when the overwhelming evidence that the known facts cannot possibly sustain establishment beliefs, forces scientists to back down and look for alternative explanations.

Should the scientific establishment be accepting any of the *alternative* theories as areas of potential research? Not necessarily — but someone, somewhere should be listening to such ideas and if and when conflicting evidence arises, it should be discussed in an open and sensibly critical way rather than dismissed as so much hot air. There is no difficulty in discovering historical precedents. Many men in the past have waged war against established opinion — Galileo is often quoted as a prime example of an early scientific heretic- but are people aware that the same war is being waged today? So many examples exist of accomplished scientific researchers who have been led into areas regarded as taboo by orthodox scientific opinion, that the term *scientific heresy* has been coined to describe their deviant theories. A *heretic* is an exponent of unorthodox opinion — and *orthodoxy* means conforming to commonly accepted opinion. The word *heresy* has all kinds of emotive baggage clinging to it, and yet a heretic is merely someone who disagrees with the establishment. They are not necessarily wrong!

There are many modern candidates for the application of the term *scientific heretic* and it has become a popular subject for discussion in the media. Some competent scientists, specialists in their own fields, became aware of conflicting evidence that certain theories relating to subjects such as homeopathy and telepathy might after all, have some basis in truth. This evidence became so overwhelming that these same scientists published their results in the mistaken belief that such information would be welcomed by their peers. In fact the reaction was the opposite of that which they hoped for. Some were demoted and others had their research grants removed. There is no deliberate conspiracy to rubbish the *heretics*, rather it is due to the training of modern scientists, the compartmentalisation of it, and the stresses and pressures that research scientists are under - pressures such as peer group pressure; specific research targets; research funding and fear of losing ones status within a very status-conscious profession. The establishment, many proudly waving the flag of scepticism, mockingly dismiss the 'conspiracy' theory, in the mistaken belief that 'alternative theorists' are claiming that there is in some dark corner of the planet, a group of people hatching plots to suppress the publication of all 'imaginative' hypotheses. Nothing could be further from the truth. The conspiracy is accidental. It is the end-product of our scientific training process, and unintentional. If the theory does not fit in with current orthodox scientific opinion then the man behind it must be wrong, or mad, or deliberately perverse. Is it not strange then, that we, the public, watch and applaud performers who claim to be able to read minds, foretell the future, and perform Uri Geller stunts, such as bending keys merely by stroking them? It almost seems that it is orthodox scientific opinion that is the *heretic* and the so-called heretics mentioned above are, along with the rest of us, of the same *orthodox* opinion.

You may, by now, suspect that I place myself among the modern *heretics*, and in a way you would be right. I have discussed the subject of this book with numerous people, both professional and lay, and it is with the former that I have received a scornful dismissal of the theories advanced in it and yet it is the very simplicity of my findings that begs for attention. I intend to show that *under certain circumstances*, Perpetual motion *of a kind,* is not only a possibility, it is an absolute necessity in the face of the falling levels of fossil fuels and the rising levels of pollution that beset our planet earth.

The foregoing statement usually creates a feeling of rising disbelief, but I would urge the reader to delay the onset of scepticism and review the evidence which I shall present. I sympathize. I had the same problem myself until, rather as some of the *heretics* described above found, the evidence finally forced me to accept what appeared to be the impossible. From an early age we are taught that perpetual motion is impossible, provable through the Laws of Conservation of energy, and of Thermodynamics. We are told that you cannot get more out of a thing than you have put in. A logical statement and perfectly true! I shall show you how, although this is true it is not actually relevant to the problem. These Laws and this saying have been ingrained in us, hammered into us, and we have laughed at accounts of early experiments to find the answer to Perpetual Motion; we find it impossible to believe that there might be a way round the problem. It is the problem of the 'closed mind'. During several years researching this book, I have come across numerous accounts of innovative thinkers frustrated by their encounters with orthodox thinkers. The journalist, Henry Lincoln, co-author of *The Holy Blood and the Grail* wrote in his book, *The Holy Place*, about the art expert, Sir Anthony Blunt, regarded as the world's foremost expert on the painter, Poussin, "I confess that I found such a closed-mind attitude utterly incomprehensible, though I have since learned that this is not an uncommon trait in the more pompously self-opinionated experts in many fields." Blunt refused even to consider the evidence brought to him by Lincoln. Further on in his book, Lincoln makes the following remark, "to say 'I don't believe' without any evidence is to make as much an act of faith as is made when one says 'I do believe'".

Some professionals have deviated from the path of orthodoxy and taken up with a heretical doctrine. Others started off on the wrong track and their claims did not become acceptable until after their death. These days most revolutionary discoveries are thought to have come from the ranks of the foremost scientists — not so! Many ideas and contrivances have been made by men who never attended university, who were gifted amateurs or became obsessed with finding a solution to a problem that was of particular relevance to them personally. Among the gifted amateurs was no lesser a person than Michael Faraday, sometimes called the father of electricity, yet he was self-taught. The obsessed 'suffer' from an undue preoccupation with an idea, and such ideas can arise out of need — there was surely never a truer saying than 'necessity is the mother of invention', and in deference to those whose work was not accepted until it was proved beyond a shadow of a doubt, as Mark Twain once said, 'A crank is only a crank until he's been proved correct'.

So what do I *really* mean by Perpetual Motion? It is known by both the lay man and the scientific community that such a device would violate one or more scientific laws. There are continuing experiments in superconductivity, in which metals or ceramics are cooled to such a low temperature that they lose all resistance to the passage of a direct electric current, allowing it to continue to flow in a super conducting ring, undiminished. This is usually referred to as 'perpetual motion of the third kind'; it is not required to do work, but merely to run continuously. But it takes enormous amounts of energy to keep the conducting material at a very low temperatures. Work continues apace to find a ceramic which will operate in the same way at room temperatures, and this may well soon happen, but the device will still be unable to *do work.* The kind of perpetual motion which we shall be looking at in this book is known as 'the first kind', and it *is* expected to *do work.*

Perpetual Motion research can be traced back thousands of years. Behind it lay a dream of free energy that could be tapped by mankind, for uses such as pumping water for irrigation or removing it from mine workings, or turning mills when the water supply dried up, or the winds were unreliable. Exactly how far back the search can be traced is restricted by the fact that we must rely on written records. In the fifth century, an ancient Sanskrit manuscript on astronomy, called *Siddhanta Ciromani* describes an attempt at a perpetual motion machine. This is unlikely to have been an isolated example even at this early period — if ***one*** person saw a need for such a device at that particular time, then others would have done so but they might not have gone into print. I suspect that if written records were available they would show that the search for perpetual motion began a long time before the fifth century, indeed there is a certain amount of evidence that pushes the date back some seven thousand years.

If a perpetual motion device were available to us today what an incredible invention that would be. If it could simply be attached to an electricity generator, there would be free heat and light for everyone. There are many possible applications, but unfortunately perpetual motion is just a dream . . . or is it?

Three hundred years ago the situation within the scientific community was no different to today's. The greatest philosophers and scientists were largely of the same opinion as modern educationalists. They deduced logically, that the chances of someone designing and building a working model of a perpetual motion machine were so remote as to be considered impossible, and this predisposed them to dismiss out of hand any such claims. For evidence, they cited the fact that no-one had ever succeeded in making one so far, even though man had been trying for hundreds perhaps thousands of years. That fact alone was sufficient proof of the impracticality of the idea, and it was made an axiom of the laws of Thermodynamics that perpetual motion machines were impossible. This axiom enabled the laws of Thermodynamics to be established. An axiom is a self-evident and undeniable truth. Now I can see that it was obvious that no record of anyone having invented a perpetual motion machine existed, (I shall call that belief into question later) and it was self-evident that no such machine existed at the time the axiom was stated (I shall call that evidence into question later) and there does not seem to be any evidence of such a machine today... but that does not prove that such a machine cannot exist; it is not self-evident and it is not an undeniable truth. But — if this were to be accepted it might prejudice the belief in the Laws of Thermodynamics, and then there could be all sorts of problems. The axiom works very well, and in fact is correct, but as Edward de Bono explained in his book, *I am Right and You are Wrong*, this is traditional 'rock' logic. Everything is defined in sharp, hard, permanent boundaries. There is no room for doubt, no flexibility, and no allowance made for unknown or unanticipated circumstances. The reality is that 'water' logic holds the correct answer. 'Water' logic is permanent but can accommodate its shape to existing or changing requirements, it will adapt its shape to that of the receptacle and it can be added to or depleted but it still remains as one volume of water. The axiom is correct but only under certain circumstances, and only when applied appropriately. But is this just word-play? I shall later show how both points of view are the correct ones!

It certainly seems to be self-evident that a machine that could run without consuming energy, would be impossible — and when you add to that the idea that it should be able to do work at the same time, it may well seem that I am wasting my time suggesting that any form of useful perpetual motion might be possible. Yet there are grounds for optimism. By the end of this book I hope that you will *know* — as I do, that a plentiful, virtually free, non-polluting source of energy for the future is under our very eyes - waiting to be discovered — or rather, re-discovered.

Our modern educational system has ruled out any chance of such energy generation, but in the Eighteenth Century, things were not so definite. Heated discussions continued as to the possibility of perpetual motion. Some believed it might be feasible under certain circumstances others, the majority of the scientific establishment, declared such an idea outrageous, ridiculous or impossible. It was known that the ultimate problem was one of friction, and work would only increase the effect that friction was already having. Heat was a consequence of friction, and the heat was readily dissipated to the surrounding cooler air, just as the Law of Thermodynamics suggested. There would never be enough energy held within the machine to draw on for more than, at best, a few hours, and it would not be able to actually do any work.

There are only two Laws of Thermodynamics, and the first one says that a certain amount of mechanical work will produce an equivalent amount of heat. In other words energy can be converted into heat, but it can't be destroyed nor created.

The second Law says that heat cannot be increased without the expenditure of more work, or energy. This means that heat can only flow out to cooler surroundings, it can't do the reverse. Today these statements seem very obvious, and when combined with the Laws of Motion as defined by Sir Isaac Newton, appear to rule out the possibility of perpetual motion.

The first Law of Motion states the rather obvious fact that a body or thing which is at rest, i.e., not moving, will continue to remain at rest, unless acted on by an external force, i.e., something pushes it. It goes on to mention that if the body is moving in a particular direction, it will continue to move in that direction unless acted upon by some external force, i.e. something causes it to change direction.

The second Law elaborates on the first Law in that it states that the action of something pushing or striking a body in motion is the same in magnitude and direction, as if it acted on the body at rest.

The third and final Law says that every action has an equal and opposite reaction — and anyone who has ever attempted to ice skate will vouch for the truth of this last statement.

So these are the laws which, in part, help to exclude the possibility of perpetual motion, and they cannot be faulted. Actually they were statements of fact that grew out of the writings of the ancients such as Thales. He was followed by Copernicus, Kepler , Galileo and Descartes. Newton combined the work of his predecessors and then added his own unique brand of genius producing his "Principia Mathematica". Remarkably, Newton himself, does not seem to have ruled out the possibility of a perpetual motion machine. It is a little known fact that in his early notebooks under the heading "Quaestiones" [sic] Newton speculates that gravity (heaviness) is caused by the descent of a subtle matter which strikes all bodies and carries them down. "Whither ye rays of gravity may bee stopped by reflecting or refracting ye, if so a perpetual motion may bee made one of these two ways." Adjacent to these words, Newton added two sketches of perpetual motion powered by the "flux of the gravitational stream". Later we shall see that he became directly involved in a famous controversy regarding a perpetual motion machine, but although challenged to comment on it maintained a dignified silence. Or was he so unsure that he dared not comment for fear of inviting the ridicule of his peers? Thus becoming a *heretic*!

I have found, in the course of my researches on this subject, that there is a tendency to invoke Newton's Laws of Motion as if they were words set in stone and immutable, inflexible and inevitable. They are all these things but Newton's words are used in an emotive sense as if one was inviting some kind of Godly wrath by suggesting that there might be a way of producing perpetual motion, and doing work at the same time, *without* contradicting the Laws of Motion.

I mentioned that the perpetual motionists were busy in the eighteenth century, but it started earlier than that, when in 1623 the King of England passed an Act, which eventually led to the establishment of the Patent Office. This allowed inventors to retain the rights to their invention and to secure a monopoly of their idea for a certain length of time. Thus they were given a chance to develop their invention and exploit it without competition. In 1635 a patent was granted for a perpetual motion machine. We know nothing about it as no description exists. In 1903, some two-hundred and sixty-eight years later, nine applications for perpetual motion-related devices were received by the British Patent Office. In those intervening years the Patent Office dealt with over six-hundred applications claiming to have designed a perpetual motion machine!

It did not end then, but perhaps due to the improvement in education of more people, and the widely held belief in the impossibility of perpetual motion, potential claimants to the secret have discovered that discretion is the best policy against a scornful public. Little is heard of the secret experiments in sheds and garages by amateur inventors still trying to solve the ancient puzzle. However a glance through the back-issues of such well-known publications as Life, Newsweek, Smithsonian, Esquire and the Science Digest will reveal articles on several modern inventors who each believe that it is just a matter of time before they solve the conundrum with a working model of perpetual motion. As accessibility improves an increasingly considerable space is taken up on the Internet, with discussions on its feasibility or not.

Arthur W.G. Ord-Hume's excellent book, *Perpetual Motion — The History of an Obsession* concludes with these comments, and I fully support them, "Talk perpetual motion for a while to the ordinary person and, sooner or later, the chances are that he will come up with a scheme of his own. Needless to say that scheme can be depended on not to work, but it proves that deep down inside us all there is a latent belief that such a thing is possible, even if for no other reason than that it would be nice to succeed where others have failed." I have included the above comments because I can vouch for its truth. Even

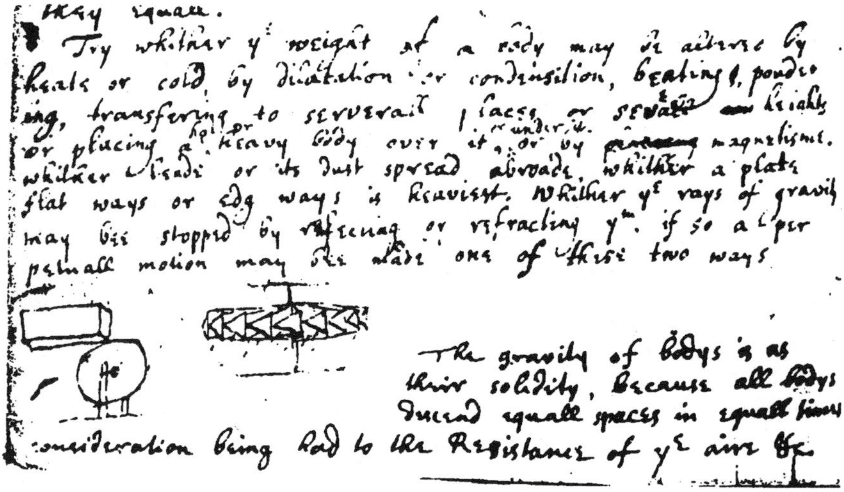

may equall.

Try whither ye weight of a body may be altered by heate or cold, by dilatation or condensation, beating, poudering, transferring to severall places or severall heights or placing a hot or heavy body over it or under it, or by magnetisme. whither Lead or its dust spread abroade. whither a plate flat ways or edg ways is heaviest. Whither ye rays of gravity may bee stopped by reflecting or refracting ym. if so a perpetuall motion may bee made one of these two ways.

The gravity of bodys is as their solidity, because all bodys descend equall spaces in equall times consideration being had to the Resistance of ye aire &c.

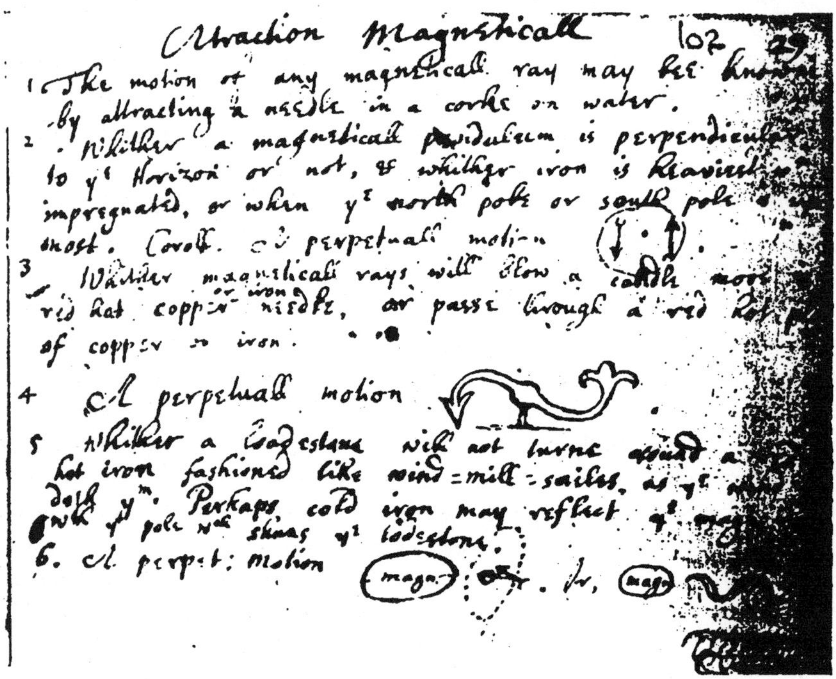

Attraction Magneticall 102 29

1 The motion of any magneticall ray may bee known by attracting a needle in a corke on water.

2 Whither a magneticall pendulum is perpendicular to ye Horizon or not, & whither iron is heaviest impregnated, or when ye north pole or south pole most. Coroll. a perpetuall motion

3 Whither magneticall rays will blow a candle ... red hot copper or iron needle, or passe through a red hot ... of copper or iron.

4 A perpetuall motion

5 Whither a loadestone will not turne ... a hot iron fashioned like wind=mill=sailes, as ye ... doth ym. Perhaps cold iron may reflect ye ... wth ye pole wch shines ye lodestone.

6. A perpet: motion mag. ... mag.

Proof that Sir Isaac Newton considered 'Perpetual Motion' a legitimate subject for scientific enquiry exists in the above notes taken from his 'Questions'. He queried "Whither ye rays of gravity may bee stopped by reflecting or refracting ym, if so a perpetual motion may bee made one of these two ways".

when the person with whom you have been discussing perpetual motion has done his best to convince you of its impossibility, and has gone away sadly shaking his head at your obtuseness, more often than not, you can expect to see him a while later with his own pet scheme.

One should not forget to add to the six hundred or so applicants who wished to be registered as discoverers of the secret of perpetual motion, a probably equal number of inventors from other countries who were besieging their own patent offices with ingenious designs. One would think that amongst this vast array of talent and optimism, that there might have been at least one successful claimant to the throne of King of perpetual motionists. But no — it seems as if no-one ever actually made a working model of perpetual motion, otherwise, where is it? No-one ever offered such a device for public exhibition. No-one ever submitted a model for testing. No-one ever offered a model for sale. But of course no such model could ever have existed because perpetual motion is a myth . . . isn't it?

Suppose one man did claim to have discovered the secret. What if he did make a working model? What if he exhibited it to the public? What if he submitted it for testing by the top scientists of the day? What if it passed every single test that could be devised? And what if he offered it for sale for twenty thousand pounds, but was unable to find a buyer who would adhere to his strict but fair conditions? What if one man did all that and eventually died in poverty, thirty-three years later, still exhibiting his machine, still hoping that someone would believe him, and still with his secret unrevealed, unsold and unsolved? Can this be possible? Could someone have succeeded where all before had failed?

Astonishingly history does record just such a man. His full name was Johann Ernst Elias Bessler, but he was better known by a pseudonym. He chose the name ORFFYREUS for reasons which will become clear.

Nearly three hundred years ago, in a small town in Germany, a most remarkable thing happened. Johann Bessler, or rather Orffyreus, as he liked to be called, had perfected his new invention, and he let the local people in to his house to see it in the hope of selling the secret to some wealthy patron for a large sum of money. The precise date was the 6th June, 1712. The town was Gera, and Orffyreus discovered a reaction to his invention that surpassed his expectations. Word spread and a crowd gathered around the device. Things got to the stage where the inventor had to close the exhibition. Later Orffyreus had the machine, which took the form of a narrow drum mounted on a pair of supports, cordoned off in an attempt to control the viewing of his prototype perpetual motion machine. Even this was insufficient; he had no desire to prevent the townsfolk seeing his wonderful wheel, but he was not prepared to risk the secret of its construction becoming known, and so, following his brother, Gottfried's advice, he began to make a small charge for admission — all of which he gave to the poor — and this did help to reduce the crowds to more manageable numbers.

The machine was not large, it was three feet in diameter (just under a metre) and four inches in depth (about ten centimetres). It was held stationary by a cord attached to the rim, and as soon as this was released, the wheel began to revolve slowly at first, but gathering speed rapidly until it was revolving at a speed of fifty revolutions a minute. It had to be forcibly stopped, but according to the inventor if allowed to, it would spin at this speed until the parts wore out. He promised that a larger machine could be constructed and this would be much more powerful.

Amazing as this was to the people of his day, it would not be a difficult trick to perform in the twentieth century, and if this was all the evidence available, there would be little point in continuing. However the reality is far from this. Over the next four years, Orffyreus, responding to ever more restrictive demands and more severe requirements that his machines perform this test or that test, produced improved wheel after improved wheel, until in 1716 he produced his biggest and most amazing device ever. I shall be going into more detail later, but a brief description may serve to pique the reader's interest.

His largest construction was in the form of a wheel or drum twelve feet in diameter (slightly under four metres); it measured about eighteen inches in depth (about forty-six centimetres) and weighed around seven hundred pounds (three hundred and seventeen kilos). It was started with a gentle push using two fingers and quickly reached a speed of twenty-six revolutions a minute, and could raise a load of bricks weighing seventy pounds (nearly thirty-two kilograms) or turn an archimedes screw for pumping water. Even more remarkable, it could with tremendous difficulty be stopped and made to turn in the opposite direction when it performed as before. The machine successfully underwent an extended test under lock and key and armed guard for a period of fifty-four days (nearly eight weeks) without stopping. To this evidence one should add the numerous smaller tests and the most thorough examinations of everything visible on the wheel, except for the internal workings which it was admitted the inventor had a right to keep secret until the desired purchase price had been agreed and handed over.

News of Orffyreus' wheel spread to the farthest corners of the civilized world. In England it met an impressive wall of disbelief, among the scientific community at least. Reports reached the ear of Peter the Great, Czar of all Russia, who went to considerable lengths to buy the secret. The perpetual motion was discussed by scientists, philosophers, statesmen, envoys and their rulers. Heated disputes arose over the fact that perpetual motion was impossible and therefore Orffyreus must be a fraud and a charlatan to make such claims. Anyway how could a mere market trader have discovered something that had eluded the finest intellects the world had ever known? However, Orffyreus was far from a mere market trader.

The general populace were curious about Orffyreus' invention. The same could not be said for the intellectuals. Letters virtually spitting venom were fired off from one scientist to another, proving that Orffyreus was a liar. Others refused to discuss the matter, their open and enquiring minds snapped shut. The most vociferous and therefore the most widely listened to, were a small group men from the town of Dresden, led by the Master Model-maker to the King of Poland. This man, Andreas Gärtner, published or had published by his followers, the most slanderous tracts and pamphlets mocking Orffyreus or challenging him to come clean and admit that his machine was a fake. There was no recourse to the law, such as a claim for libel, so there was little Orffyreus could do other than go into print to defend himself. This he did with great enthusiasm, pouring out on paper all his injured hurt and pride.

After the success of his first wheel, the small one at Gera, Orffyreus constructed a larger version, which measured five feet (1.5 Metres) in diameter and turned as the others had done, without stopping. Wonderful as this machine was, it still attracted criticism. "It was still too small"; or "a larger one would be more useful"; or "a larger one would not work, because as everyone knows the larger a machine is, the less efficient it becomes." These were the kind of comments made about the wheel. The most telling one in the opinion of the inventor was the published statement by his enemy Gärtner, that the machine was wound up, and that was where its motive power came from. Orffyreus went away to ponder and came back with an irrefutable argument; he made a wheel that could turn in either direction, thus obviating the possibility of a clockwork mechanism.

This did not silence his enemies; enemies because Gärtner had been joined by Christian Wagner, a student of mathematics, whose published tracts were, if anything even more vitriolic than Gärtner's. Wagner tried to claim that Orffyreus had fitted some kind of split axle to his wheel in order to allow it to change direction. He does not explain how this might be achieved, and becomes something of a target for Orffyreus' sarcasm. The new machine was submitted for examination and passed all the tests. As usual these results were insufficient to satisfy those who believed that the inventor was a fraud, and he left town yet again after first smashing his wheel to pieces in disgust.

News of the inventor's work and the remarkable claims made for it spread, and the Landgrave of Hesse-Kassel offered to take him and his machine and granted him his patronage. He made Orffyreus a Commercial Councillor and invited him to rebuild his wheel, bigger and better than ever before. This he did, but in spite of the successful tests, his enemies would not leave him alone and to Orffyreus' dismay the detractors eventually won the day, and in fact, from the moment he first demonstrated his invention, a storm of controversy which had swiftly established itself, raged about him. When he died thirty-three years later, alone, in poverty, and still claiming that the invention was genuine, his own, and never having been proved a liar, that storm, that hurricane of controversy had not abated in the slightest.

The legend of Orffyreus' wheel is fairly well known, but by omission and distortion of the facts, the real story has never been told. Most of the information which appears in every encyclopaedia in what ever language, tends to originate from one particular source. In 1795, fifty years after Orffyreus' death, Friedrich Wilhelm Strieder published a biographical dictionary which included an account of the inventor's life. In it, Strieder mentions that he has used Orffyreus' own published writings as a source of information, complaining at the same time that he has needed "great patience — I must say — since it is a truly abominable piece of prose". In fact the account is written with some style and considerable wit and humour, but is overly long and repetitive when complaining about his enemies. I can understand that one might need patience to read all of it — Orffyreus' writing does suffer from an excess of verbiage — but it is not abominable.

The veracity of Strieder's account of Orffyreus' life can be checked, fortunately. There are many Orffyrean tracts extant in which he describes his life, so the two versions can be checked for discrepancies. It is fairly obvious that Strieder has a very poor opinion of the inventor, and does not hesitate to let us know on several occasions, but I have not been able to find any glaring difference of a factual nature between the two accounts. There is, however, a tendency to be selective in the use of certain facts. Strieder lays greater emphasis on some facets of the story, whilst playing down other aspects.

Another source for accounts of the actual devices exhibited by Orffyreus are the newspapers of the day, and they provide much information about both the dimensions of the *wheels*, as they were called, and the prevailing opinion regarding the possibility of perpetual motion. However the best mine of information lies in the numerous letters written about the Orffyrean machine. Two of the most striking and convincing letters came to my attention through the efforts of William Kendrick (1725?-1779), a rather pathetic character, who made a living penning literary articles. Kendrick actually died while waiting for the Patent Office to grant him a patent on his own design of perpetual motion. Whether it existed outside his own imagination is difficult to say, but certainly he was a believer in the possibility, and indeed gave a course of lectures on the subject.

But to return to Strieder, his account continues from the time of the big Kassel wheel, albeit sketchily, up to the inventor's death. Unfortunately we only have Orffyreus' own account of his life up to and including the building of the last wheel. It is, therefore, harder to determine how much faith to place in Strieder's account of Orffyreus' life after 1719, because we have nothing to compare it with, except what we can piece together from the numerous letters and newspaper reports. It is reasonable to assume that it was a factual rendition of events, but biased. The reason for this bias is not hard to find. In his search for information Strieder turned up an extremely damning piece of evidence. Namely, no fewer than seven sworn statements to the effect that the inventor was a swindler, and that they, the witnesses, had been forced to take part in the fraud, and had been sworn to silence by Orffyreus. Most of the witnesses were either relatives of Orffyreus or his servants.

This evidence has been the overwhelming factor in the demise of Orffyreus' reputation, let alone the fact that perpetual motion was, even at that time, regarded as a scientific impossibility. Faced with these facts the reader may well feel that it is pointless to even suggest that there could be anything other than fraud connected with the Orffyrean machine. All is not as it seems, however. The biographer only reproduces two documents as examples, and as if they were not bad enough, he summarizes some of the accusations from some of the other documents. He says, for instance, that Orffyreus' "wife confirmed in great shame that she was forced to swear five oaths to her husband — who always carried consecrated wafers, but was a desperate man who cared for nothing — and that, in the end, she nevertheless felt obliged to confide secretly in a man who was bound by oath and duty, but that that man decided to take her treacherous husband's side and do nothing, even warning her of the most frightful consequences of any further action".

Can there possibly be any doubt of Orffyreus' guilt? History does not think so, and neither did I; at least I wouldn't have done except for one small seemingly insignificant phrase which cast doubt on the authenticity of the evidence. The full versions from both documents will be reproduced later, but for now I want to extract a small piece of text from the affidavit sworn by Orffyreus' maid.

In a document dated 28th November, 1727, Orffyreus' maid makes the following statement, as recorded by Strieder:

"The posts had been hollowed out and contained a long thin piece of iron with a barb at the bottom which was attached to the shaft journal. Turning was carried out from Orffyreus' bedroom which was close to the machine, on a shelf behind the bed."

Strangely, it was these particular sentences, which Strieder had quoted as one of the most vital pieces of evidence of the inventor's duplicity, that eventually led me to write this book. I had read the account of Orffyreus' wheel before, and accepted regretfully, that the inventor was a fake. It was not until I came across the legend again much later, that I first had my suspicions aroused. I re-read the story, and was left with a feeling of unease. I knew that perpetual motion was impossible, but there was one glaring error in the account which would give me no peace; and in addition there was another loose end which needed tying up.

The first area of concern was the statement that the maid made concerning the secret mechanism. It was frankly, impossible. Whatever mechanism moved that wheel, there is absolutely no way that it could be driven by the means described. Twelve feet across or 3.6 metres — eighteen inches thick or 45.7 centimetres and weighing an estimated 700 lbs or 318 Kilograms, and the whole construction turned on a pair of bearings measuring just three-quarters of an inch! And what is more, accelerated from a very slow speed to one of between twenty-five and twenty-six revolutions per minute, in just three revolutions. The maid says that the posts were hollowed out and a barbed piece of iron inserted and connected to the shaft journal. Anyone giving reasonable consideration to this account of the maid's will see that the power and strength required to keep a machine of this size turning, by applying its force *through the bearings*, would be enormous — and besides where would one find a metal of sufficient strength to withstand the tremendous load placed on it. Not only that — it had to be kept turning for *almost eight weeks — and* was expected do work; raise a box of bricks weighing seventy pounds and turn an Archimedean screw. Forget the problem of actually devising a mechanism which would operate inside a wooden post connected to a bearing at one end and a small wheel at the other!

It is hard for me to convey the utter certainty in my mind, of malicious intent by the maid. I am sure that whatever criminal accusations may be laid at Orffyreus' door, this account of the maid's is fictitious. And if this part of Strieder's evidence, the key part, is faulty, how much reliability may be placed on the rest of his evidence? But I mentioned an additional loose end. During Orffyreus' efforts to achieve recognition for his perpetual motion machine, his claims came to the attention of Karl, the Landgrave of Hesse-Kassel. He was, perhaps, similar to a Duke in England, and as such he ruled a small Dukedom, whose capital was the town of Kassel. It is recorded that he was the only man that Orffyreus ever allowed to see inside the wheel. He was sworn to secrecy, and he never broke his word. It puzzled me to think that if he had seen the inside of the wheel, he obviously knew the truth, and yet no mention was made in Strieder's biography of any comment by Karl, following Orffyreus' arrest for fraud. It seemed to me that I needed to find out more about this man. He took Orffyreus and his machine under his patronage, and even made the inventor Counsellor of Commerce for Hesse-Kassel. Could he have been a party to the subterfuge? Or might he have been duped by a plausible rogue? Or — the only other possibility — was Orffyreus telling the truth? Had he really invented some kind of self-propelled machine?

Having cast doubt on the accusations repeated in Strieder's biography, I was left with a problem. How on earth *did* Orffyreus make his machine move? It could not have been perpetual motion, because I had to accept the fact that that had been declared impossible. Was it fraud? I could try to find out as much as possible about Karl. Perhaps there would be evidence that he was a man of doubtful character, or alternatively he might have been naive, and easily taken in. Perhaps some of the witnesses to the action of the wheel were unsuitable to judge whether what they were seeing was genuine. Maybe they were bribed? If it wasn't fraud, how was the wheel driven? Could there possibly be an answer which did not require the overturning of well-established scientific principles? But the first area for research was the inventor himself. Why did he claim this secret for himself? Why did he ask what was an enormous sum of money for those days? (Equivalent to over one and a half million pounds or nearly two and a half million dollars today!) How did he expect to get away with the money if his machine was a fake? If it was a fake, how was it done? If all these questions could be resolved without blame attaching to either Karl or his protégé I would be left with only one problem — if it wasn't perpetual motion, what was it?

The main source for details about the early life of Orffyreus is his own pamphlet which he called his *Poetica Apologia.* The piece was written during 1716 and 1717, and is in the form of rhyming couplets. It was composed at the same time as he was building his twelve foot wheel at Hesse-Kassel. The building of the wheel was a considerable undertaking for a man working virtually on his own, yet he was able to carry out his job as Counsellor of Commerce, and also compose and set down some seven and a half thousand lines of poetry describing his life from birth to date.

I had intended to include the complete translation of this work of Orffyreus', but it has proved to be too bulky. I wanted to show Orffyreus' writing because it conveys his sincerity more effectively than anything I could say, despite being translated from 18thC. German into 20thC English prose. The original is in simple rhyme, but in the translation, it is as if Orffyreus was here talking in front of us about his life, his problems and his successes. Either way I think it is a worth-while document so I shall tell his story as he wanted it told and use many quotations to convey the spirit of the man. The single, most abiding impression left after reading his *Apologia*, is that Orffyreus had no desire to defraud anyone. All he wanted was to be taken seriously and treated fairly. That he believed in his machine is beyond doubt. But was he deceived?

Map of Orffyreus' travels from when he invented his Perptuum Mobile in 1712 to his death in 1745.

Chapter Two

ORFFYREUS

Orffyreus was born Johann Ernst Elias Bessler, but prior to his discovery of the secret of perpetual motion, he adopted the pseudonym by which he is known today. Although the date of his birth is not recorded, Strieder located a document which must have been originally in Orffyreus' possession, which has a note on it, in the inventor's own hand. This informs us that he was in the town of Lauban on the 1st May, 1696 when 122 houses were burnt to the ground in a single night. He says he was fifteen at the time, so he must have been born in 1680 or '81. This document also contained a map of the Oberlausiz area. Written across the map is the following inscription.;

"Regina Fuchsen, nee Maucken, the sister of Orffyreus' mother, lived and died in Jonsdorf near Zittau and is buried in Bertsdorff (not far from there)."

Orffyreus went to school in Zittau so, since his aunt lived in this area, he probably came from one of the forest villages near Zittau, which was on the border of Bohemia. He says his parents were of "good, honest, pious stock", and he was their eldest son.

One of Orffyreus' endearing traits was his unembarrassed self-assessment of his personal qualities; thus he describes his parents' "great pride and joy in him" and their realization "even while he was still in his first flush of youth, four and a half to be precise, that he was already showing much promise — promise which he fulfilled to a large extent." He describes how his father made a small toy wheel for him to play with and "this was his particular favourite". Clearly he had a busy and enquiring mind, and he was sent off to the school at Zittau, where "the poor mite learnt a lot and quickly too." After a short while the Rector of Zittau, Herr Weise, took personal charge of his upbringing, "caring for him as if he were his own father".

Herr Weise pushed his young charge hard and he responded by "each day asking questions about new skills, and taking every opportunity to acquire knowledge". He was quiet, seldom laughing and he "kept his own counsel". His natural curiosity "brought him to high seats of learning, and in the day to day world, graciously granted him many friends, books, good food, and much money". However, inevitably he learnt the "bad ways of the world, neglecting his duties and studying false teachings". He fell under the influence of "an evil man and began to associate with people who held no respect for Christian teaching. Duelling, dancing and riding suited him well, but God punished him by causing him to wear out his shoes of evil along rough paths of illness". Orffyreus returned to his studies, "shunned all vanities . . . sparing no pains to become a wise man", and "avoided all revelry, ate only bread and soup and drank only water; refrained from feasting, gambling and carousing". He strove to meet men of wisdom and "developed every opportunity to increase" his knowledge of "the workings of nature". His efforts eventually led to him becoming "successively a Cleric, a Doctor and finally a Mathematician". Despite these claims, he was probably an amateur Cleric, Doctor and Mathematician, although that does not necessarily reflect on his prowess in these fields. Everything that he wished to master, he laboured at with great thoroughness, in order to become well-versed not just in theory but in practice also.

He studied mechanics, and ruthlessly picked the brains of anyone who might be of use. He "toiled endlessly at his creations, sculpting, grinding glass, smelting, casting and turning". Whenever he met a problem he would turn to the experts to seek help, and would pretend that he "could hardly count up to three", in order to allay their suspicions as to his motives. Over a number of years he acquired skills in making clocks, glass-blowing, painting pictures and copper-engraving. He learned the art of enamelling and wax-modelling. He became knowledgeable about astronomy, and the surveying of woods, meadows and fields. He worked with gunpowder and constructed guns, claiming that he was a good shot himself. He studied music and learned to play some instruments.

He began to earn a lot of money and soon acquired many friends, but as he grew more successful and boastful and arrogant, he also "became very aggressive to anyone who so much as looked at me!" He got into fights and "although it never came to daggers drawn, he did get quite badly shaken up once or twice". Once again his lifestyle shamed him into "taking communion again".

During his travels he had to be rescued on two occasions from bands of marauding soldiers, who sought to press-gang him into the army. Orffyreus may have been working as an army medic because he says that for a while he had to give up medicine because he became afraid at the sound of gunfire.

Fortunately he escaped their clutches with the help of a nun who "let me in one door of the convent and out the other".

The following day he met and rescued a man who had "fallen into a pond whilst trying to collect water in two huge ewers. He accompanied him to his home having dragged him out on to dry land". The man was an Apothecary and taught the eager Orffyreus many things including "true medicinal arts, not the quackery he had dealt in previously". He learned several "cures" and also how to "display a proper air of sympathy to his patients, so that he could make money out of them"!

Having learned much, and feeling the wanderlust again, he was persuaded to travel with a "fine" gentleman who intended visiting foreign parts. During their travels they visited a monastery and it was here that Orffyreus first conceived the idea of searching for the secret of perpetual motion. "I happened to notice a spit turning, and immediately fell to wondering how it could carry on doing so, seemingly all by itself. It had the power to continuously rotate, like a clock that could wind itself up. I studied carefully and took note of its skilful construction."

Orffyreus began to associate with craftsmen, in order to learn their skills, and as he had picked up a smattering of other languages he was able to converse with people of divers nationalities. Always he was curious, asking questions, trying his hand at different trades, always seeking new ideas, looking for the clue that might put him on the track of the secret that he now sought.

Eventually he arrived in Prague, and there he met a Jesuit priest — "perhaps the most learned priest I had ever met" and they became friends. Orffyreus discussed his ideas with the priest, who joked that a perpetual motion machine with legs on would be a wonderful discovery. They joined forces but the priest suggested they enlist the aid of an elderly Rabbi who lived in the Hebrew quarter of the town. He says that he has "been seen too often recently going into the Hebrew quarter, creeping along to see some Rabbi or other". He warns Orffyreus that it does not take the Jesuit brothers long to "sniff such things out", and asks him if he will visit the Rabbi on his behalf. The Priest wrote a letter of introduction to the Rabbi and Orffyreus took it along with him to his first of several meetings.

The Rabbi "had a great beard just like the prophets of old". Having asked him several searching questions, the Rabbi then swore Orffyreus to secrecy, saying that he was to be the intermediary between the two priests. As was his custom, Orffyreus extracted as much knowledge as he could from the old man, learning something of Hebrew, and ancient Hieroglyphics, "the language of nature and the writings of angels". He studied and ran errands but had no success with his perpetual motion, "it seemed no wheel would turn for" them. The "Jesuit, by now old, ill and weak, gradually lost interest as he realised he had not long to live. He passed on to me a lot of his things. One day the two of us bade each other farewell, and I wished him eternal joy". Once again Orffyreus set off on his wanderings.

He left Prague and travelled to Dresden in the company of a miller's son. His reason, he says, was so that during their journey to Dresden, when they called into several mills, Orffyreus was able to pretend that he was an apprentice miller and picked up their milling jargon. In the process he learned much about their techniques such as whetting millstones and replacing their bearings, and many other practical tips that might bear fruit later. He says that " the remarkable inventiveness of the master-millers was reward enough for him".

Once in Dresden Orffyreus wasted no time in seeking out additional information or techniques that could be of use to him in his search. Using the same method as before, he befriended a carpenter's apprentice. "Over a few glasses of wine" he put his idea to the apprentice and soon Orffyreus was passing himself off as a carpenter's apprentice. His true motive, as ever, was to learn the secrets of each trade that he encountered.

Within a few days he left to resume his travels again, passing himself of as a preacher this time and selling pamphlets. He says that this took place in 1703, which makes him twenty-two or twenty-three years of age. He was widely-travelled and could obviously look after himself by this stage, and although it is impossible to say when he left school, it was probably around the age of fifteen, so he had been on the road for about eight years. He makes no mention of having returned home, although he may have done, and if he did, I do not believe he would have stayed long — throughout his life he displayed the same nomadic tendency.

Now Orffyreus' wanderlust took him further afield, this time through Holland, England and Ireland. On his journey he met an Irishman who had some ideas on the subject of perpetual motion, but Orffyreus dismisses them as "irrelevant", and in fact began to hold his own ideas in higher esteem. However, things were getting a bit desperate, money was short, so Orffyreus took up medicine again. At this time he began to work on a book or notebook, in which he noted down everything that he had learnt. Included in it were designs for rotating wheels, pulleys and various devices that occurred to him from time to time. All things mechanical were entered into his book, and he slept with it under his pillow at night.

On his return to the mainland of Europe, as he was crossing the channel, he fell ill, and some of the crew threw the book overboard along with most of his

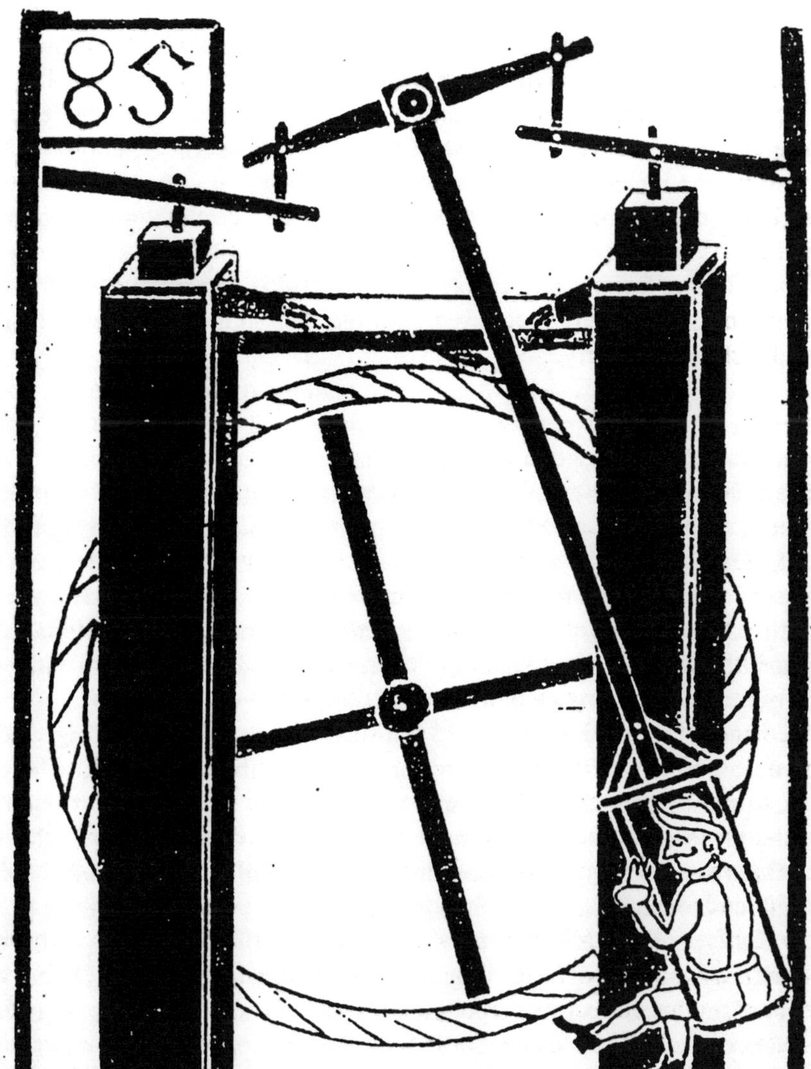

The delightful engraving reproduced above, is taken from Orffyreus' "Maschinen Tractate", and shows us that the inventor also had a sense of humour.

belongings. Perhaps they thought he would be unable to pay his way if he became sick. This did not dismay him overly. He set to, making a few pocket-watches and "other such trinkets to earn a bit of money — who could have been richer than I?" In a newspaper report he happened to read that " a thing to be prized more than a ton of gold would be the invention of a Wheel which could turn of its own accord". This spurred him on, and soon he could think of nothing but the secret of perpetual motion, "hoping that through such a device I might become seriously rich". Now he decided to come home, feeling that he had found his true vocation. All was clear and nothing could go wrong. He searched for and found a workshop where he could carry out his experiments. He worked industriously with great optimism, but nothing came of it and he grew despondent again.

On one of his journeys he heard of a rumour of buried treasure. Some treasure hunters that he met, were so enthusiastic and guileless that he resolved to join them. He had made the acquaintance of some "eminent people" who were sponsoring the hunt, to the extent that they had wagered their entire houses in partnership with the treasure-hunters. The law was brought in to scrutinise the operation and Orffyreus was convinced that his fortune was secured. They all began to dig, but the hole filled up with water and they ended up with nothing. At this stage they all turned on Orffyreus, roundly abusing him, and blaming him for their lack of success. He rounded on them with equal ferocity, returning the insults with enthusiasm.

Once again Orffyreus returned to the straight and narrow, asking God for forgiveness. He settled down again, studying his subject and making experiments, but without success. Through this period he suffered depression. He considered the pupils who had been at school with him and noted that although he had shown more promise than they, he was still unsettled and they had married and obtained good positions in life. "The years were piling up one after another, and I still had not got my own little fireside in proper order. I couldn't see such a thing happening to me for a long time, if ever. Others were rich in possessions, but I was a poor soul who seemed to get closer to ruin the richer I became in knowledge. Yet I took comfort from the constant presence of God within my heart". He forced the negative thoughts from his mind and concentrated on his goal. If any man could discover the secret which he sought then that man was he. He told himself that he knew so much about so many diverse crafts, he was in the perfect position to find the solution. He must put aside his cares and concentrate on his task and perhaps, he thought to himself, "God will allow me to fathom the mystery of true motive power".

Following this time of doubt, Orffyreus attacked the problem with renewed vigour and intensity. He reviewed all the designs that he had dismissed to see if there was any way of improving them. He acquired new skills and improved the old ones, he consulted old books by people who had written on the subject of perpetual motion — and there were many. And perhaps most importantly, sensing a kind of revelation in the air, Orffyreus began to take an interest in the church-organ. He had encountered these splendid instruments many times in his travels and had come to appreciate their subtle mechanisms. The more he discovered about them, the more he wanted to learn.

To aid him in his work he turned to a relative who was an artist, organ-maker and turner. The skills that this man possessed would surely be of use, so Orffyreus approached him, and took him into his confidence, knowing he could trust him. A large organ began to take shape under his assistant's guiding hand, but Orffyreus' real motive was to build a perpetual motion machine. The organ-building was both a cover and a means of examining the many complex systems involved in this art, which reached its apogee at this time, particularly in Germany. He was certain that in the art of building organs, lay the secret. Somewhere amongst all the different stops and registers, chords and key-boards lay a clue. Among the sliders, rods, trundles, levers, stop-knobs, groove-valves, stop-lever bars, stickers, backfall bridges, trackers, fulcrums, roller arms and bearing shafts he would find the inspiration necessary to achieve his goal?

Orffyreus was right. The organ of that period was probably the largest and most complex machine ever constructed to that time. Orffyreus set to work with a will. He made many calculations and created numerous models. He bought in parts from other manufacturers, and when all was ready, fitted them together with painstaking care. When the moment came, and the wheel should have begun to turn freely . . . it didn't! Over and over again, the same result each time. The wheel refused to budge. Orffyreus would explode with anger and frustration. But he would not give up. He complained to nobody, but doggedly gnawed at the problem. The cost of constantly buying material for his prototypes swallowed up their meagre funds, and Orffyreus ate sparingly of only the plainest foods. Sometimes a brilliant idea would come to him in a flash of inspiration, and he'd work feverishly to get his idea translated into a working model, only to find that it remained steadfastly motionless. At this result he would throw himself on the bed and "weep bitterly, till on one occasion, God took pity on me, showed me his love, and blessed with good fortune my ensuing labours".

Nothing happened immediately, but having fallen asleep, exhausted, he had a strangely invigorating dream, which gave him renewed confidence to carry on. For weeks he forgot the outside world and concentrated on nothing else but the solution to his puzzle. He slept in his workshop; he worked all the daylight hours, and long into the night by candlelight. He ignored all interruptions, and then at last — the enlightenment he sought finally materialised. A new design that he had created in his mind was transformed into reality — a real perpetual motion machine which moved spontaneously! The problem that had been troubling his mind for so long was solved — and having found the solution, all sense of urgency left him. Many things that he had neglected now surfaced in his mind, demanding his belated attention.

Having received two letters from "two rich patients, I stopped all work, took the coach that had been sent for me, and rushed off, leaving my work in pieces, but with the thought that it didn't matter now that my trust in God had proved so well founded". He cured both patients who were generously appreciative, and through them got to treat other sick people, "who I was likewise able to cure". Orffyreus now travelled to the Erz Mountains. He visited water-mills and mines in the area, because he needed to know if and how his newly discovered perpetual motion machine might be of use in such occupations - whether the machine could be built large enough to cope with the loads imposed on it, and how to connect such a device. He arrived at Annaberg-Bucholz where, in order to earn sufficient funds to maintain his way of life he continued to practice medicine.

The town mayor and also chief physician was Dr. Christian Schumann, and Bessler came to know him. Bessler obviously respected Dr. Schumann, and says he was a man of great medical experience, however this may have been a case of him blowing his own trumpet, for the following reason. The mayor had a daughter, his youngest, who was lying seriously ill. "She was screaming, tossing to and fro, and twisting her limbs. Her parents were greatly distressed, and had tried, to no avail, many ways of alleviating her suffering. I gazed at her and said that it was awful that such a lovely girl should have to suffer so. Straight away I set to with God's help, and within the hour I had cured her. As a reward I was given the girl herself! With God's grace I took her hand in marriage". This was in 1711, and Bessler was already about thirty years of age.

A fairy tale ending! Orffyreus was ecstatic. He knew exactly what he wanted now, and the sense of urgency that had left him for a short while, returned. He wanted to get back to his home and develop his new machine and market it. He set off with his bride — and her maid. They were all blissfully ignorant of the major part she was to play in the inventor's downfall. Before the day was out the travellers ran into great danger. "We suffered storm, hail, thunder and lightning, sleet and rain! We were soaked — but worse, the coach began to slip down the hillside, and to the screams of the women, the water rushed up to meet us. We feared we would drown; the people on the hillside could only watch in terror, as the coach ended up in the river. My wife was almost overcome with dread. Crying out to God to help us, I tried to comfort her. God's miracle saved us - he did not send us into the abyss, but for a whole hour let the coach float along until we eventually reached the bank.". Orffyreus ordered the coachman, "Upwards, away from the river, as fast as you can!" They eventually arrived home and Orffyreus settled down to the married life that he had looked forward to.

They set up home in the town of Gera, the same town that was to witness the first exhibition of the Orffyrean perpetual motion. Initially they lived at his cousin's house, and it was here that Orffyreus began to construct his new machine. However he found that none of the rooms in the house were big enough for what he had in mind, so he took a lease on another house which had much larger rooms. The problem was that living with his cousin, Orffyreus worried that his secret might be accidentally revealed — it would be difficult to work without risking someone coming in seeing something he did not want them to see. As well, he would have needed space to work, and it may have been crowded with two families living together. It is probable that this was just a temporary arrangement until the newly-weds could afford something more permanent.

The leased house was owned by Herr Richter and stood on Niclaus Hill, and the inventor actually constructed his first demonstration model here in the year 1712, exhibiting it on the 6th June. The device stood three, to three and a half feet high (a little over one metre). I should point out here that all measurements are approximate because each town had different systems of measurement regarding both weight and distance. The wheel "was mounted in such a fashion that anyone could walk all the way around it. On unfastening a bolt, the wheel immediately began to revolve. Whoever wished to push it or touch it at any point was free to do so. People could see there was no trickery involved".

News of the inventor's machine spread quickly. "People were flocking to my machine. First people of the highest rank, then the minor aristocracy, scholars, clerics and other learned people from all over". The scholars argued over the news. Some were for Orffyreus, others stated that it was impossible. A "decision was arrived at; the machine should be scrutinised in minute detail. Should the thing be true, it was said, then it would be of great potential value, and the wide world should be informed of it. If, however, the thing was a fraud from the outset, the perpetrator should be prosecuted! Well this was done, and I received a certificate testifying to the genuineness of my device".

This might have been sufficient, but it wasn't. "Nit-pickers came along. They said though my machine was a work of true artistry, it was merely a curio, which could never be of practical value, and would remain a mere toy. But if I could build one five Ells high, then my claims would be believed. (Five Ells is between nine and ten feet, or nearly three metres). "Soon I was being plagued by crowds of Sunday visitors, some of whom were fairly rough and boisterous, and I couldn't even eat in peace. I made a firm resolve that as long as I was in control of matters, no wheel would ever turn an inch on Sunday. Certain enemies of mine got to hear of this, and began to congregate outside in ever greater numbers, shouting and jeering. It got so bad that finally, in a rage, I smashed the machine into a thousand pieces, and vowed to seek peace and quiet elsewhere".

This was not what Orffyreus had anticipated. He believed that he would be welcomed as a hero, shouldered on high and carried through streets full of cheering crowds. He was not disheartened though, he "soon found all I was looking for at a place called Draschwitz. Here I constructed a five Ell high wheel, and anyone who wished to could come and see it, revolving between two narrow beams. The bolts which regulated the motion were screwed into and out of the axle by many people, for I allowed my friends to operate it. When the facts became known, many of the newspapers ran articles, and soon I was getting visits from Royalty, Nobility and Academics".

As before the house rapidly began to fill, "bursting with spectators, some foolish, some cunning". Orffyreus found that he had more than his share of those whose sole motivations were grudges born of hatred and envy. While all this was going on he continued to treat the sick successfully, and this too caused comment. Some suggested that the wheel contained a dog or a cat, because of the scratching noise which came from within. But at the top of Orffyreus' list of dislikes were those who alleged that he was a wizard, a hangman's apprentice or a follower of Dr. Faust — perhaps he had imprisoned a genie in the wheel — and worst of all, since he was having so much success with the sick, he must have signed a pact with Death himself. As a deeply religious man he took extreme offence at such suggestions. Still the determination to continue with his project remained strong in him. He comforted himself by thinking of "all those past great inventors who came up with so many useful devices, and only after they'd died did it become the fashion to praise them, for during their lifetimes they were snarled at by the ignorant, who couldn't achieve anything at all". His thoughts were remarkably prophetic.

Things were proceeding little better at Draschwitz than they had at Gera, however the inventor says he "was lucky to be treated well at Draschwitz", at least he was dealing with "persons of high rank and not foolish children". He complained that some people were heaping too much praise on him, that the honour was not his, but God's. Some people were asking many questions and not listening to the answers. Wild speculation and rumour added to the confusion, and amongst those involved were a couple of noblemen from Dresden. Through them word got to the ears of Herr Andreas Gärtner, soon to be a major thorn in Orffyreus' side.

Gärtner sent Orffyreus a letter asking such questions as, how long will the device run continuously, without being driven by the kind of machinery used by millers, clock makers, and carters? Orffyreus replied indignantly, "your questions can be interpreted as suggesting that my invention will not move unless it is driven by water, weights, oxen, wind or even manpower". This was an offensive letter if Orffyreus was genuine, implying, as it did, that there was trickery involved. The inventor's answer was simple, "as long as the Lord shall wish it," and "if Herr Gärtner would come and observe the machine in action, watching quietly in an unprejudiced fashion", then he, Orffyreus would answer any questions and dispel any doubts.

Letters were exchanged, but the Master-Modeller to the elector of Saxony still had doubts.

"Will any obligation be required from those who wish to inspect the device?"

Orffyreus' answer was "no, none."

"How long will he be permitted to observe? "

"As long as is wished, the machine will continue no matter how long you stay."

" Can it be used for raising weights?"

"Yes, it can raise a weight, placed twelve inches away from its axle, once every second."

"Is the machine designed to be moveable?"

"Yes — If anyone wishes to purchase it, he can depart with it, and it with him!"

Orffyreus was not the most tactful of men, and the flippancy displayed in these answers to someone as important as Gärtner, was not guaranteed to ensure everlasting friendship. The Model Master was an old man and owed some respect due not only to his age and position, but also to his own skills displayed impressively, in the form of complex clocks and other mechanical devices, to most of the monarchs in Europe at this time. He was not going to have his position eroded by some uneducated peasant from the forests, with a self-proclaimed perpetual motion device which everyone knew was impossible.

Not long after this, the inventor fell seriously ill. Gradually he recovered but dissatisfaction with the way things were progressing effected him so much so, that Orffyreus decided to move again. An important Court offered to put lodgings at his disposal, but he had grown disillusioned with the area he was living in, and rejected their offer. Subsequently another court made an even better offer, but he rejected that offer also — a decision which he admitted was a mistake, and which he blamed his friends for. They promised to find a suitable place for him move to, but as he despondently says, "somehow it all went wrong, and finally having to leave Draschwitz, I smashed my wheel to pieces".

That winter, Orffyreus was forced to take a home in the tiny village of Obergreisslau. He did not like living there, and as he was out of the public eye for a while rumours began to spread that he had run away. He suffered one of his recurring bouts of depression at this time, and found it hard to motivate himself, following his illness. However as Spring arrived he started to think about his predicament again. The rooms he was living in were too small, and just after Easter he moved to Merseburg, a village not far from Leipzig. He says that people wondered why he liked living in villages, but he said he could not live in towns, preferring "fresh air and open spaces. Climb up to my attic window, look out and see for yourself! That's the pleasure I get from living in the beautiful countryside".

Once again he set to work on a new machine, this time building one six Ells in diameter (over 11 feet or 3.5 metres). This one was to a new design, internally. It could be set in motion in either direction, and the reasons for this departure from the original design will quickly become obvious. "It caused me a few headaches before I got the mechanism properly adjusted. Why did I make this wheel, you may well ask? I shall tell you. During my stay in Obergreisslau my detractors put out the cunning falsehood, that my device, like a clock, needed winding up. This caused me to make some changes to the mechanism so that intelligent people would appreciate the falseness of such a proposition."

Despite this new device, people still tried to suggest that the basis of its power lay in the machine's winding mechanism. Orffyreus' exasperated answer to them was, "*you* go and make a wheel that turns *both* ways, before you start with your accusations again!" — He goes on to demand to know where the doubters were when a local newspaper advertised the fact that the inventor was happy for anyone to come along and observe his machine in ceaseless rotation day and night. As he said, "nobody came to see it, and so only Nobody has seen it, and since the inventor's enemies won't take Nobody's word for it, the doubters remain to be convinced".

Time was passing and Orffyreus was no further forward in selling his machine than he had been after the first examination. He complained that he had to answer the same questions over and over, several times a day. Some visitors would try to sneak a look at the machine's interior, while their friends kept him distracted with questions. Groups of foreigners arrived, "clambering all over the tables and work-benches, until I packed them off to the nearest hostelry. They complained that the whole thing was a disappointment and they should have been allowed to inspect the interior of the machine. Some of them even tried to rip off the cladding on the wheel in an effort to see inside". To those who pretended that his secrets were already common knowledge, and

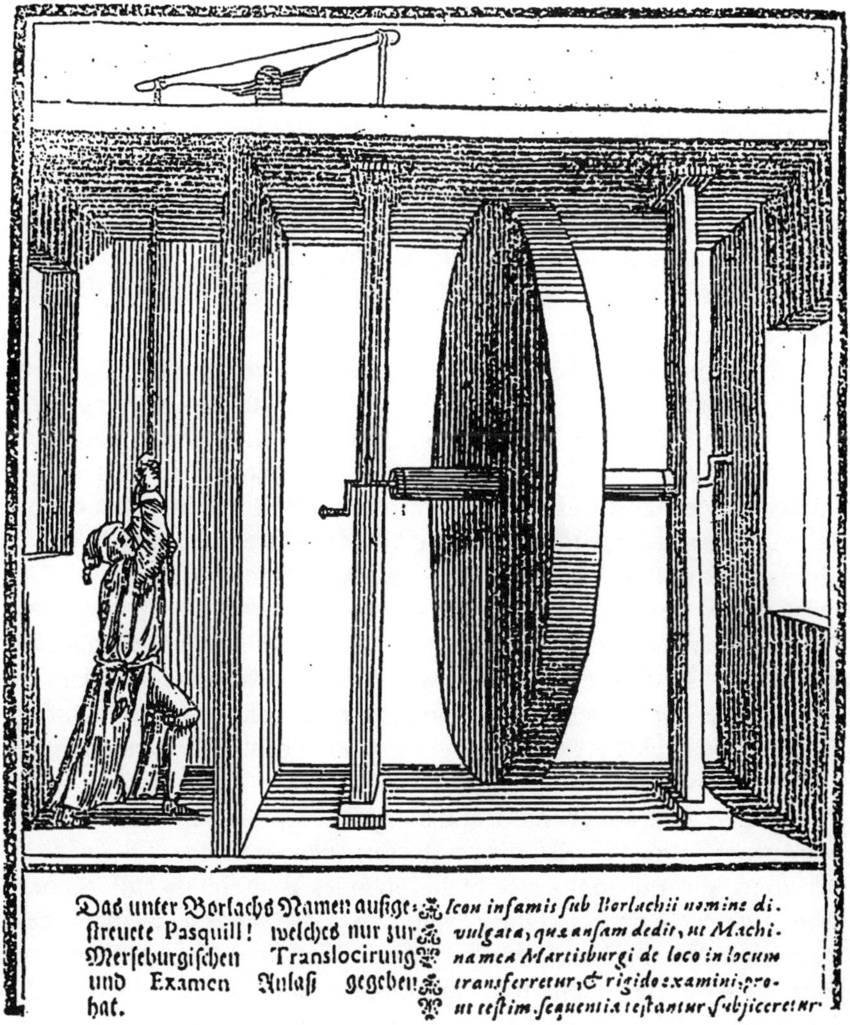

This cartoon is the infamous 'Borlach' satirical engraving which Bessler contemptuously reproduced in his 1719 publication – 'Das Triumphirende Perpetuum Mobile . . . '

therefore there was no need for him to be so secretive, Orffyreus would dress up his explanations with mumbo-jumbo. He would mix in made-up latin words, and then clam up tight, and people would whisper that he had worn himself out with excessive study. "Never did I hear a word of praise, though I was criticised from all directions. But I never changed, never once wavered, because a single word could have betrayed my wondrous achievement.".

During that summer, Orffyreus chanced to bump his head very badly in a fall. "As I lay on the ground, people thought I'd never get up again. I crawled around on the ground like a worm. When they got me into bed I felt as if I no longer had a head. Eventually I recovered, with God's help, and was dismayed to learn that Herr Gärtner had secretly been to my house while I was still in bed very early on the 22nd July, coming from Dresden with Borlach the satirist". Borlach was another of Orffyreus' protagonists, and was accompanied by Christian Wagner. "Upset by the strangeness of it all, my assistant, Bessler, showed them my wheel. Upon which Gärtner and his two accomplices promptly slipped away again. Soon all three were talking about me and making up lies about my invention, and how I had deceived the whole world. My life's work was held up to ridicule. They made a copper-engraving for this purpose in Dresden".

This cartoon purports to illustrate how Orffyreus is supposed to make his machine turn. The drawing shows the wheel suspended on its two pillars. Next to the machine, but inside a secret room, is a tired-looking servant, hauling on a piece of rope, which is connected via a rocking beam mechanism to a rope hidden within one of the supporting posts. This drawing has been cited over and over again as proof of fraud, and completely ignores the evidence of the various examinations carried out on the Orffyrean machine.

Orffyreus' reaction to the visit by his enemies as he rightly calls them, was to protest at the unfairness of their examination. He suggested a more reasonable approach would have been to contact him first and ask for a proper appointment, during which all his questions could have been answered and any tests he wished to carry out could have been accommodated. This, the inventor maintained, would have been the action of an honourable man. Not so in Herr Gärtner's case, and therefore an opportunity had been lost to have his claims vindicated by a man of great importance. "People now know", he continued, "that he was largely responsible for the untruths put out about me, and only added Borlach's name because he was afraid to stand in the firing line alone. He has lied to the world, and has used an exaggerated acquaintance with me to attempt to show the world that I am a charlatan who acts only from motives of self-aggrandizement".

Orffyreus, wounded and suffering the pangs of self-righteous indignation at the wrong he perceives has been done to him, cannot leave the slur on his good name unanswered. Pages and pages of his *Apologia* are devoted to the insinuations of his enemies, Gärtner, Wagner and Borlach. Over and over again he demands to know what he has done to deserve such attacks. Repeatedly he asks them to come and look at his machine, examine it, test it and pass a fair opinion on it — or shut up. If he is lying, have him arrested, chop of his head, but first give his machine a fair trial. He tells them to make a similar machine which can do all or more than his machine, and if they then admit that their's is a fake *then* he will give up and acknowledge that they are right and his invention is inferior. Of Wagner he says, "His poisonous rantings bite like the mad hounds of hell. His main theme is to try to force the world, through the power of his words, to believe that my device can't run for a single day, and that it isn't what I claim it to be, and that it is like all things that have to be wound up".

In answer to other suggestions he says, "who wants to sit beside a machine for perhaps three or even six months? I get letters ad nauseam from people who claim that I should observe such machines for lengthy periods, but if those people were to come and see my machine running for many weeks they still would not be satisfied, unless I revealed the secret to them".

One of Wagner's argument relies on the proposition that the wheel is driven by a roasting-jack, such as Orffyreus saw in the monastery when he first conceived the idea of a perpetual motion machine. The tract in which this argument appears, contains a diagram of a roasting-jack, and Orffyreus scornfully also reproduces this in one of his later publications, along with Borlach's insulting diagram. Wagner suggests that, like a roasting-jack, the wheel will slow down to a stop in a very short time, and that "even a weakling could stop it just by holding the axle". Again the official tests call the lie on this claim. Wagner goes on to accuse Orffyreus of being an accomplished swindler who won't engage in a serious conversation. He recalls numerous conversations with the inventor during which he is supposed to have made many outrageous claims for his machines. Orffyreus replies, "When did I ever give you any information about my machine? When did you ever call on me at Merseburg? When were you in my machine-room there?"

It seems that Wagner paid Orffyreus just one visit while he resided at Draschwitz. "I received him warmly, but our acquaintance lasted just as long as it took him to swallow his breakfast. I hadn't seen him before and I haven't seem him since, and yet he spouts forth all manner of false rumours, as if he knew all about my craftsmanship from long years of unbroken conversation with me. The only purpose of his labours is to destroy me and my work utterly".

Another of Wagner's stratagems involved getting Gärtner to take action

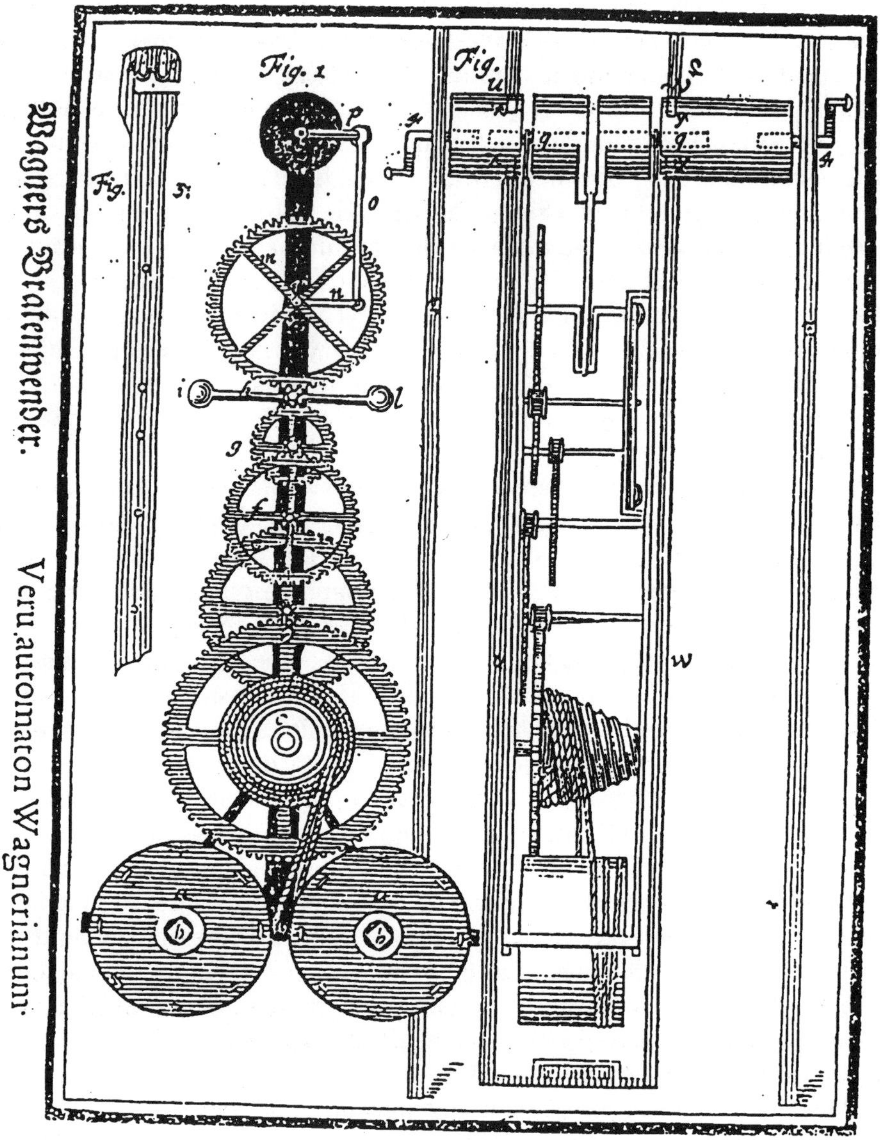

This is a copy of Christian Wagner's "solution" to the Orffyrean machine, also reproduced dismissively in the 1719 publication – Das Triumphirende Mobile . . . It is a common roasting-jack used in large houses in the eighteenth century.

against him. A libellous letter came into Orffyreus' hands containing tales of alleged action by Gärtner who it seemed had had to seek redress in court for sins committed by the inventor against him. Gärtner, it was said, had had to paint a verbal picture of Orffyreus which was so defamatory that it could not be printed. The inventor contemptuously dismisses such stories as lies and rounds on his attackers in picturesque language. "Begone you evil hornets, you prattlers swollen with poison, you envious wretches who have so vilely debased me! Only your lap-dogs will be enticed by your putrid crumbs".

All this acrimony came to an end in August 1716, when Karl the reigning Landgrave of Hesse, took the bitter Orffyreus under his protection and gave him lodgings in his castle at Kassel, and a job as Counsellor of Commerce. There was of course a price to pay. Karl was not going to be conned by a smart fraudster. Before he allowed the inventor to take up residence, he had to agree to a condition. Karl must be allowed to see the inner workings of the machine. This must have cost Orffyreus dearly. He was adamant that no man should know his secret unless he had paid the asking price, £20,000 at that time, around one and a half million pounds at today's rates, or well over two million dollars. Karl was no fool. He persuaded Orffyreus that he would swear on oath never ever to reveal what he had seen, until the inventor had profited satisfactorily from his invention. He agreed to finance the building of the new machine, and as a final sweetner, he agreed to pay a sum of 4000 thalers to the inventor as a fee for gaining temporary access to the machine's innermost workings. Reluctantly Orffyreus agreed.

Once he had made his decision, he could not sing Karl's praises highly enough. To the end of Karl's life he sent him panegyrics on his birthday, and while not wishing to debase Orffyreus' motives, one cannot help wondering if all this fulsome praise for Karl did not originate from some degree of self-interest on the inventor's part. After all, if he fell out with the Landgrave, it was possible that despite the oath, Karl might reveal details of the machine's design , whether by accident or on purpose. He needn't have worried. Karl never spoke of what he had seen, except to dismiss rumours of deception that followed the inventor's arrest later.

While he was negotiating with Karl's emissary, he received numerous letters from interested people. One of these contained several questions which the inventor sets down and answers in his Apologia. The first question asked if anyone of eminence had seen inside the machine. Orffyreus' answer was no. Karl had not seen the inside at that time, but the question may have predisposed Orffyreus to consider the point — if he could find someone whom he could completely trust, and who was independently wealthy and therefore beyond bribery . . . Another question asked if a larger machine could be constructed. Orffyreus suggested that with help, it should be possible to construct a machine over 20 Ells in diameter, (35 –40 feet, or 10–12 metres). Yet another question asked how much such a machine would cost to build. The answer — anything from 100-400 thalers (£20-£80 at 1717 rates). The same questioner wanted to know how long it would take to build the device. The answer, said Orffyreus depended on how many men were involved. What had taken him six months to do, could be done in four weeks with several assistants. Would the machine be durable? It depends on what materials are used, came the reply. If the main parts are constructed out of the best iron and steel and brass, then of course it will be more durable than a less costly construction. Will the machine work in the cold or will it seize up. The answer — "yes one can go on working comfortably in winter with my wheel singing away merrily. When other machines are snowed up, mine will go on and on. Hard frosts may reign over there — but here summer lightning flashes!". Finally, the writer wanted to know, would he sell his machine to one man or many. Orffyreus answered that he only wanted the one-off payment. Once he had got that, he wanted the whole world to share his machine, "so that I might be able to see its fruits rise up everywhere in towns, hills and meadows. No village shall find the cost too great."

One of the letters he received, asked Orffyreus how he could be so certain that his machine was inspired by God, rather than the Devil. Surely, the writer asked, if God was behind the revelation, would not someone have discovered it more than a thousand years ago? And if the machine became well-known in many different places, wouldn't it place a large army of people out of work? Orffyreus' answer was that he was in a way unique. Many people had searched for the answer — clever people with special talents — but that they had all been in various offices, positions or professions, and were unable to devote their entire lives, to the exclusion of *everything* else, to the search for the secret of perpetual motion. The odd hour here and there was insufficient for "Mobile research". His determination not to give up, had been a crucial factor, but also it needed the practical skills that he had developed, not just the theoretical ability. This had uniquely suited him for the task he had set himself. It was only after he had constructed over a hundred machines over a period of ten years that he found success.

The inventor foresaw many uses for his machine. He visualized it driving water mills in hilly country, when the water had dried up in times of drought, thus providing continued employment. "Another use that people will quickly find for my machine is in pumping out mine-workings that have become flooded with underground water; too often people who would like to earn a little money for themselves have had to sit idly back while the potential riches of such sites go to waste because of excessive water". He also considered that there will be applications in smelting and stamping. "I also envisage the use of my machine in pleasure-gardens. Water could be brought from distant ponds to create fountains, and little ornamental water-features with fish for the enjoyment of gentle-folk".

"So why," he asks, "does anyone wish to trouble themselves asking pessimistic questions, saying my mobile is a device which will bring misfortune to poor people? Many thousands of people will get work, who are now forced to live on spiders. All mills will be safe, as will other establishments, for my invention will be incorporated in places where no water is available, such as the Erzeberg region". The Erzeberg area is and was an area of mining and particularly suitable for a water-pumping machine for use in the many mines there.

All this sounds perfectly reasonable, but one of the complications was due to Orffyreus' demand that no access to the interior of the wheel would be granted until all the money agreed had been paid over. He says," It ought to be made perfectly clear to everyone that there is no possibility of anyone being cheated in this matter. When all the agreed money has been paid in full, all the secrets of its artistry will be revealed. Nothing will be hidden away. Afterwards the money will be mine and the secret the buyer's. Of course, if my work should turn out to be as my enemies describe it in their ceaseless rumour-mongering, (its motive power is not inherent — it has to be wound up — it does not run for as long as its materials endure etc) then the buyer can snatch back his money, and have my head chopped off! I shall then deservedly be covered in shame, derision and scorn. And all my enemies shall be rewarded. They will clap and cheer with delight".

It is hard to imagine that Orffyreus would really have gone to the above lengths unless he really believed in his machine. He goes on to tell us that if nothing comes of all this, and the world continues to scorn him, he will smash the wheel to pieces and make his living as a medical practitioner.

I mentioned in the first chapter that the inventor was persuaded to make an admission charge in order to try to reduce the crowds attempting to see his machine. He tells us about it in response to a letter asking him why he smashed his wheel, just as people were beginning to ridicule his enemies. The occasion was at Merseburg. His second wheel was attracting large crowds. A lot of "common, undesirable rough elements got in who didn't understand a thing about it". Some of his close friends suggested that, as he had been keeping a vow to give money to the poor each week for some ten years, it might help both him in his aid to the poor, *and* help to reduce the crowds, if he put a poor-box in front of those who came to view the machine, so that they could make a voluntary contribution. Orffyreus agreed with the suggestion and had his assistants make a collection, "and the crowds diminished sharply!" He claims that as people wanted to know how much money the poor were now getting, he revealed that they were receiving much more than had been collected, so he was still giving money from his own income. Despite this, he removed the collection box.

What forced this action, was a letter from Herr Gärtner, in 1715, in which he reproached Orffyreus for taking this action for his own benefit. Not long after this, in January 1716, a decree was issued in Dresden, to sequestrate each day a sum of six groschen, (about £2 in today's money). When the inventor was faced with having to pay a tax on money he was collecting for the poor, he decided, he would rather give up and start again elsewhere, out of his enemy's reach. It is clear that Gärtner and his friends were behind this shabby trick, aimed at forcing Orffyreus into poverty, and his machine out of public view. They all lived in Dresden, and were kept fully informed of events in the town of. Orffyreus comments that they no longer attacked his machine, but aimed their hatred at him directly. He became so frustrated at his enemy's action that he smashed his wheel and moved on, as we saw, to Obergreisslau.

Now, however, having successfully negotiated a deal with Karl of Hesse-Kassel, he moved with his family to Kassel. The standard of the arts, the safety of its citizens, the goodwill, support and advice he found there made a lasting impression on Orffyreus. At the head of this small but powerful state stood Karl, and he was everything that Orffyreus sought in a prince. He ruled with a strong hand, but was fair. He had a profound interest in the latest scientific discoveries, and he had an open mind.

At this point we shall depart from Orffyreus' own account. About to enter the service of Karl, things looked rosy for him. Karl had guaranteed his protection for Orffyreus, and decreed that no man should speak ill of the inventor, lest he endure the wrath of the Landgrave. He even extended his guardianship beyond the boundaries of the state of Hesse, in a clear message to Gärtner and his friends. That this proved effective is evidenced by the deafening silence from Orffyreus' enemies during the time he spent under Karl's protection. It is clear from his autobiography that Orffyreus' account of his search for the secret of perpetual motion is genuine. While it is acknowledged by everyone that perpetual motion *is* impossible, Orffyreus definitely *thought* that he had discovered it. Was he deceived? Or had he discovered a new form of generator?

To dig deeper into the mystery of Orffyreus' wheel we need to access a number of letters written about the inventor at this time. More can be ascertained by investigating the numerous newspaper reports of the tests carried out on the machine. Most importantly what sort of men carried out the examinations? Could they be trusted as impartial witnesses? What of Orffyreus? Can we extract any more information about him from the people who knew him? How were the machines tested? Were the tests as tough as they could be. If all reasonable doubt about the validity of the tests can be removed, and those tests are found to be as rigorous as possible, then we are left with the conclusion that, like it or not, Orffyreus was not a fraud.

Chapter Three

THE EXAMINATION AT GERA

Orffyreus wrote his *Apologia*, seven thousand lines in length, while he was building his mighty twelve foot wheel. This task would probably have been carried out in the evening, by candle-light and with a quill pen. He must have written an average of nearly forty lines per night, if he wrote every night for six months, so it was no mean faculty of his, to have written this 'truly abominable' piece of poetry, as Strieder put it. Much of the credit for drawing out this particular talent of Orffyreus', must go to his former headmaster, Christian Weise.

Weise (1642-1708), is chiefly known as a novelist and playwright, but was also respected as an administrator. He was born, lived most of his life and died in Zittau, where Orffyreus obtained his education. While attending the University of Leipzig, it is said that Weise wrote poems to order, often turning out ten or more daily — a talent that he seems to have encouraged in Orffyreus. While headmaster at Zittau, Weise wrote plays for his pupils which were known for their realism. They were based on sound reasoning as the fount of knowledge and the actors were encouraged to use dialects. The language was said to be often crude or violent. He instilled in his charges a practical approach to life which would aid them in making their way in the world. His ambition was to produce educated business men and courtiers for the service of the State. Training in eloquence was paramount.

Christian Weise had a profound effect on Orffyreus, and the latter tried to make use of the skills handed down to him by his teacher. As Weise had taught him to, he sought employment within an office of the State, finally succeeding at Kassel as Commercial Counsellor. He then attempted to further fulfil his teacher's aspirations for him by writing his biography in simple rhyme. Having accepted his teacher's precepts, Orffyreus wished to carry out the requirements expected of him, but was repeatedly deflected from his primary aim by his need to earn great fame and fortune. Weise's constant encouragement in the practicalities of life and business are apparent in the actions of the inventor. Learning the crafts and functions of every conceivable trade, but not settling on any one particular art to the exclusion of others, was not exactly what Weise had in mind, but Orffyreus' search for the secret of perpetual motion offers a motive and an explanation.

Later Orffyreus published pamphlets in which he sought to unify the Christian religions; his goal, following the successful sale of his perpetual motion machine, was to build a school for teaching all the various crafts available in those days. This was to include teaching of the Christian religion in a general sense without any of the dogma associated with different sects and denominations.Many of the ideas promoted by Orffyreus stemmed from the teachings of Christian Weise.Weise's father had been exiled for his Protestant beliefs, and Weise himself had absorbed much of his father's protestations. The Rector of Zittau school taught his children to be pragmatists in religion; to be understanding of other beliefs; and to follow that ideal that they felt comfortable with. Some of the violent outbursts by Orffyreus when describing his enemy, Wagner, can no doubt be attributed to the training Weise gave his students. Orffyreus was a favourite of Weise and he was an accomplished public speaker and a natural actor. This ability coupled with his thirst for knowledge and his talent for practical subjects helped him to success in his vocation.

In 1712, on the 6th June, Orffyreus had exhibited his perpetual motion machine for the very first time. After years of hard slog and drudgery he had succeeded where all before had failed. He was on the verge of fame and fortune, but how to possess those pinnacles of ambition? His plan was simple and had been worked out during the long years of research. Now the machine was proven to *his* satisfaction at least, he could literally set the wheels in motion. The mechanism driving his wheel was very simple - once the secret was out, anyone would be able to copy it. So he had to devise a foolproof method of obtaining his reward. He would only have one chance, as the first buyer might reveal the secret. Even if he managed to keep the method secret for a week or a year, sooner or later someone would get hold of the information, and before long everyone would know. The only way forward was to have a one-off payment of sufficient size to enable Orffyreus to retire, and to carry out his long-cherished ambition of a Christian non-denominational craft-training school for peoples of all nationalities.

The asking price for the secret of the perpetual motion machine was £20,000, an enormous sum of money in those days. The inventor surmised that only kings, princes or other rulers were likely to pay that amount of money to have the machine. It would seem that Orffyreus had an uphill task to secure such a deal, and we shall see that he did have great difficulty, but not in finding potential purchasers — no, the problem lay in the actual negotiations; the buyer needed to see the workings of the machine to satisfy himself that he was not being cheated - but the seller would not allow it in case *he* was cheated of his reward. There were no patent laws available to protect his invention, none with any teeth, so the course he had set himself was the only one open to him, if he wished to get a worthwhile sum of money in order to fulfil his future plans.

Because of the very simplicity of the mechanism, Orffyreus was constantly on guard against a chance remark which might have given the game away. Despite his care, Orffyreus gave away several clues as to how his machine worked, and this taken together with the numerous letters, newspaper reports and certificates gives us a good insight into the workings of his machine.

The certificates were issued by various local authorities and from their content it would seem that the tests were carried out in increasingly exemplary fashion. No sane person would go to the court of a ruling monarch and ask for his most trusted and cleverest advisors, culled from the cream of the universities, to attend at a certain place on a particular day and test his machine for fraud — not unless he was absolutely confident that his machine was genuine. Even the head of that court, a Duke, was invited to observe the tests, and he agreed. It is inconceivable that anyone would take such a risk, especially when considering what had happened to previous 'entrepreneurs'. In his *Apologia*, Orffyreus states on more than one occasion that if he is revealed as a fraudster and his wheel is not what he claims, then he must forfeit his life by having his head chopped off. If that seems an unlikely consequence, one only has to consider the unfortunate end of Edward Kelley. He was an English alchemist and a companion of the famous astrologer, alchemist and mathematician John Dee. In 1595 having travelled to Prague to see King Rudolf's personal alchemist's laboratory, he lost his life by having his head chopped off. The court of Rudolf II of Prague took a dim view of the fact that Kelley's attempts to turn base metals into gold had failed repeatedly. He was not the only man to lose his life through attempted fraud.

Another alchemist from England also entered the service of Rudolf II, taking his wife and children with him, although he fared better than some of his predecessors. His visit to Prague took place in 1610 and coincided with a visit by the above-mentioned John Dee. His name was Cornelius Drebble, and in fact he was born in the Alkmaar, Netherlands in 1572. He lived and worked in England from 1605 until his death in London in 1633, apart from his visit to Prague. He published a book in Holland in 1605 entitled, 'Of the Nature of the Elements — The Wonderful Discovery of Perpetual Motion which the Philosopher of Alkmaar has Contrived'. He came to England at the wish of James I. He was required to conduct scientific experiments and to make scientific toys for the amusement of the King. An entry in the diary of Hans von Vrenden, secretary to Frederick, Duke of Wurtemburg, dated 1st May, 1610, says that 'his excellency visited Eltham Palace to see the perpetual motion; the inventor's name was Cornelius Drebble, a native of Alkmaar, a very fair and handsome man, and a very gentle manner'.

Orffyreus spent some time in Prague around 1700, and doubtless heard accounts of the activities of the alchemists and the harsh treatment meted out to those who tried to deceive their employers. He may well have heard of the experiments of Cornelius Drebble. Rudolf II, King of Prague and his predecessor, Maximilian II were both Holy Roman Emperors and made Prague the 'metropolis of alchemy', sponsoring ever-hopefully, most of the leading alchemists in Europe. The reputation of Prague as a centre of excellence for the study of alchemy was still strong in 1700 when Orffyreus visited the city. Taken together these factors tell us two things of vital interest to him. The first thing was that rulers were interested in scientific toys and would gladly pay for the privilege of possessing such items, sponsoring the development of them, and paying the inventor great riches if successful. The second thing was that those same rulers would react with extreme violence to anyone who sought to defraud them. Punishment was severe and immediate; no powerful ruler could afford to let anyone get away with blatant deception. An example had to be made to deter others from similar adventures.

After the first rush of excitement at Gera, the inventor needed to have some kind of official recognition, so he requested and obtained the first of many official tests of his machine. He believed that a certificate clearly stating that his device had been inspected for signs of intent to deceive, and had been cleared of that objection, would be sufficient. Also in the certificate, the tests carried out to divine the power and potential of the wheel would be included. Having received his certificate, Orffyreus waited for the buyers to come.

He describes how he ran to the Dowager-Countess, and although her name is not mentioned my research indicates that she was of the house of Reusse. Countess Reuss was referred to in the Certificate as, 'Her Grace Countess-Widow and His Grace Count XXII'. She was therefore the Count's mother. Orffyreus knew the Count of Reuss in his capacity as a Doctor of Medicine. The two patients who sought his services just after he had succeeded in his endeavours to construct a perpetual motion machine, came from the Reuss family. He effected a cure in that family, and this explains his presence at the

Count's table. It is likely that following the success of his search he welcomed the letters from the Reuss family, as they would have been the ideal aristocratic connection for him to request an official test of his new machine. The Countess seems to have taken a liking to the inventor so she persuaded her son to make the arrangements for the test.

Despite the success of the test, it was four months before the inventor received his certificate, and they must have been nail-biting months for him. The certificate is brief and tells us little of what actually took place, and this fact was obviously brought home to Orffyreus at some stage, because the later ones go into considerable detail about the tests carried out. The first test was presided over by Dr. Johann Georg Pertsch, Sr. (1651-1718), 'poet laureat'. The value of the president as a practical scientist is doubtful, but we have no reason to doubt his trustworthiness as an impartial observer. The test took place on the 12th October, 1712 and was witnessed by various local notables. The certificate reads thus:-

> 'The long-sought after and desired Perpetuum Mobile has been invented and constructed recently, through God's grace, here in Gera. It is a unique and highly useful machine that rotates without any external weights, wind, water or spring mechanisms. It has its own quite extraordinary internal Motus Perpetuum that not only maintains, moves and turns it round continuously, but also it is able to drive easily other machines for which a great force is necessary, for example water works, mills and similar. It has become quite famous, against expectations in all neighbouring towns. In addition, news has come that so-called friends intend to steal, at the earliest opportunity, the secret of this machine in order to construct it themselves. They have already begun to attempt to secure their rights to it in another manner.
>
> 'Herewith, it is certified that Dr. Johann Ernst Elias Orffyreus, Medical Practitioner and Mathematician, born in Upper Lusatia, has been living here for some time and has behaved in an honest, Christian and modest way. According to his reputation, he has deserved only praise and honesty.
>
> 'Through God's grace, he is the originator and constructor of this unique masterpiece, and there is no one else who is entitled to lay claim to have invented it. The machine was constructed for the first time by the above-mentioned Mr Orffyreus, here on Niclaus Hill in Mr Richter's house, and put into its remarkable speedy rotation on the sixth of June of this current year 1712.
>
> 'It was shown several times to Her Grace the Dowager-Countess

> and His Grace the Count in person; and also, by their gracious consent and agreement, everything was shown and made available to us. Now we have undersigned this Testimonial in confirmation of the above, and to attest to the fame and honour of the inventor, and for his special protection, benefit, recommendation and promotion. It has been issued for all persons high and low, who love Art and Knowledge. It has been signed and sealed by our own hand at Gera, 9th October, 1712.'

The certificate ends with a list of the signatories, fourteen in all, plus the phrase

> '...and many others who were there.'

The certificate, reproduced above is fine as far as it goes. Unfortunately it does not describe the tests done, nor for how long the machine ran. It relies largely on the belief that these were men of unimpeachable reputation and were therefore beyond criticism. Reading between the lines, Johann Adam Cass, a surveyor and mechanician, was chosen to carry out the tests and supervise the actual examination. He was clearly an expert as far as such a title could be applied at that time. He would have been able to detect any signs of deception, although it could have been possible for such a small model to have been driven by clockwork. Orffyreus' landlord, Rienhold Richter was present as were one or two mathematicians and teachers. The rest of the court came with the Count and Countess of Reuss to Mr Richter's house on Niclaus Hill to inspect the new machine. To suspect that Orffyreus would subject himself and his machine to such an august gathering while secretly planning to deceive all of them, including his new found aristocratic friends seems doubtful.

The certificate could be used as a kind of 'letters patent', indicating the inventor's prior claim to the discovery, and possibly affording him some protection against other claimants if such came forward. This idea had worked in England, for instance when Newcomen and Savery were developing their steam engine. The British Government provided some ten years of protection against their competitors, and in fact extended it at a later date, before the patent ran out. It is doubtful if Orffyreus' certificate would have given much protection. The court of Reuss was a minor aristocracy with no real power and no teeth to enforce any protection offered. At the time, however, Orffyreus was reasonably pleased with the outcome of the tests. He said that the certificate 'testified to the genuiness of my device. I've still got it at home, by the way; anyone can see it if they so desire'. He concludes by saying that there was no more talk of his machine not being a true mobile.

That Gärtner resented Orffyreus is beyond doubt, but why is harder to fathom. It may be that he was convinced that perpetual motion was impossible, so anyone claiming to have the secret *must* be lying. At one time Gärtner had become involved in a bet with the Court Jeweller, J.F. Dinglinger, who believed that he could make a perpetual motion machine. The machine never materialised so Gärtner won the bet. It is clear from this and other evidence that Gärtner was a firm believer in the impossibility of perpetual motion. What ever the reason for the dispute, Gärtner and his two supporters, Borlach and Wagner became a thorn in Orffyreus' side for the next three years, and again after Karl had released Orffyreus from his service, patronage and protection.

In *Apologia*, the inventor tells us of two castles which were offered to him as a base from which to develop his machine. One was probably that of the Reuss family, but as he had left the nearby town of Gera in considerable dudgeon, he declined their offer. The other was a 'worthy Saxon Duke, no less' who offered his home to Orffyreus. This time the inventor blames his friends who persuaded him to decline the offer. The Duke was Moritz-Wilhelm, Duke of Zeitz, a cousin of Karl of Hesse-Kassel, and a regular correspondent of Leibniz.

Gottfried Wilhelm Leibniz was Germany's leading philosopher, scientist, mathematician and historian. He was a true genius and communicated with all the most famous and accomplished men in Europe, over six hundred! At Moritz-Wilhelm's court were several learned men with whom Leibniz also corresponded, chiefly on the subject of Orffyreus' wheel. Leibniz's interests were boundless. Besides being interested in science and mathematics, he was also a lawyer, a historian and a political adviser. He is best remembered for his invention of the differential and integral calculus. His dispute with Sir Isaac Newton over who invented the calculus first will come into this story later, but at the time we are concerned with he was officially researching the history of the Guelph family, of which George of Hanover, soon to be George I, King of Great Britain following Queen Anne's death, was the head. As if Leibniz had time on his hands, he was also involved in developing a mechanical calculating machine, which could not only multiply and divide, but could also raise to a higher power and extract roots. This invention led to his being admitted to the Royal Society. On top of this, Leibniz had a longstanding interest in the technical problems of mining. This interest led to revolutionary findings in geology and power economy. It is impossible to cover the depth and breadth of Leibniz's interests in a short paragraph, and the reader is recommended to read further biographical material. Most if not all biographies will, however, probably omit all reference to Orffyreus for the simple reason that perpetual motion is and was regarded as impossible and therefore any interest in such claims shown by Leibniz was regarded as a bizarre temporary aberration, and not worthy of inclusion. It was something of a surprise to me, therefore when I discovered that he wrote many letters enquiring about the invention, and was able to visit Orffyreus and examine the machine on at least two occasions. Leibniz was fascinated by the idea that Orffyreus might have discovered something of value to the mining industry, even though, he assured his correspondents, what ever was the source of the power within Orffyreus' wheel, it could not be perpetual motion. He, it was, who suggested a more rigorous form of testing of the wheel which Orffyreus subsequently arranged.

Although Leibniz died on the 14th November, 1716, in the short period of his acquaintance with Orffyreus' invention he sat at the centre of a network of written communications providing information about the wheel, which spread throughout Europe. He discussed the Orffyrean machine with Peter the Great, Czar of Russia and several leading scientists of the time so that it became common knowledge.

Peter the Great, a magnificent innovator who was attempting the heroic feat of dragging the vast feudal country of Russia into the eighteenth century, was intensely curious about all things western. In spite of his size, said to be around seven feet, his giant body seemed almost incapable of containing his enormous, boundless energy. He worked harder, played harder and lived faster than any of the people who he came into contact with and frequently became irritated with those around him who were unable to keep up with his phenomenal workrate. His particular interest was in the sea, but his curiosity about the latest technology and recent scientific theories and achievements was comparable. He was already discussing with Leibniz the educational reforms that he intended for his people, and the establishing of an academy of sciences to direct education and research. There was considerable correspondence between Peter, Leibniz and Areskine, Peter's personal physician. In 1712, Peter appointed Leibniz a councillor of justice in the Russian service with the special function of advising on educational, legal and governmental matters and promised him an annual salary.

In the eighteenth century Leibniz's name was inevitably linked to that of Johann Christian Wolff (1679–1754). Their philosophy is referred to as the Leibnitzio-Wolffian. Wolff was a brilliant and popular university professor, with continuous literary and academic success. He graduated at Leipzig in 1703, where he attracted the attention of Leibniz. His first post was professor of mathematics at Halle University. He also lectured on Physics and later Philosophy. At the time of his death he was Chancellor of Halle University, Privy Councillor of Prussia and a Baron of the Holy Roman Empire.

He was a pupil of Leibniz, and it was Leibniz who recommended him for the mathematical professorship at Halle University. The reasons for his inclusions here is that he was also a scientific adviser to Peter the Great, and helped to found the St. Petersburg Academy of Sciences in Russia. He too was a great communicator and wrote several times on the subject of Orffyreus' wheel to Peter's representatives, and most importantly he was present at one of the examinations of Orffyreus' wheel, and reported to Leibniz all that occurred during that examination.

In 1716, Wolff had published a 'Mathematical Lexicon' in which he discussed the possibility of perpetual motion, saying,

'Such a motion has been sought since ancient times, in vain however, sometimes with considerable expense... but those who are busy trying to find perpetual motion become perpetual fools".

'It should not be held against great mathematicians, if they choose not to waste their time with this investigation, because this invention is regarded as such that its finding depends not just on intellect but more on good luck. For example, if we take just one obstacle into account; parts of the machine rub together during motion and we are unable at present to calculate the friction. Also there is no sufficient theory to explain other aspects such as what form the construction of such a machine should take. So those who seek perpetual motion cannot avoid exhaustive toil and expense in order to create a machine that has some chance of success. It would be wrong to expect those who use their intellect to promote knowledge, to diversify into the unknown in order to achieve something by so much toil and expense. In any case an opinion has been established for a long time now, that perpetual motion is impossible.'

Wolff goes on to quote other early scientists who set out to prove that perpetual motion is impossible. Finally he records that Leibniz, his mentor and friend, has satisfied himself that perpetual motion is impossible. Leibniz wrote that,

'No research should be made into such things [as perpetual motion]. That way we shall retain the tranquillity of mind and our souls shall rest in peace. So no respected mathematician will seek purely mechanical perpetual motion (the impossibility of which has been demonstrated), and no prudent man should tire himself in vain experimentation. And what is more, no one should trouble other people with this, or incite unnecessary aggravation in people who are not in agreement with them. He should exert his intellectual powers in areas in which he can succeed.'

It is clear then that the climate of intellectual opinion was against Orffyreus from the start, not only denying the possibility of perpetual motion, but actively discouraging people against researching the subject — and even from discussing the subject with others, in case they disagreed.

The case *for* perpetual motion is looking *very* bleak, but all is not lost — a way can be found to accommodate the views of conventional science, as well as those who think that there might be something in Orffyreus' machine. To get at the truth we must look at some previously unpublished letters which throw new light on the workings of the mighty Orffyrean device. Some have been published in part, but always any reference to Orffyreus has been excised. These letters have been ignored for several reasons, and looking kindly on their omission, one might suggest that the subject of perpetual motion having been effectively expunged from orthodox science, there was no requirement to include the extracts containing references to the Orffyrean wheel. The letters make fascinating reading and allow us to share the excitement and urgency apparent in the news that Orffyreus' wheel generates. The letters do not reflect badly on the famous scientists concerned, but rather, show a creditable open-mindedness that modern-day scientist could do worse than strive to emulate.

Chapter Four

LEIBNIZ AND WOLFF

After leaving Gera for Draschwitz, Orffyreus came within the domain of Moritz-Wilhelm, Duke of Zeitz, and thus the news of his wheel came to the ears of the great Leibniz.

Moritz-Wilhelm (1664-1718), was a distant cousin of Karl, Landgrave of Hesse-Kassel, and as they ruled over neighbouring territories, they were on familiar terms with each other. One of Karl's sons married Moritz-Wilhelm's daughter, Dorothea-Wilhelmina. The Duke of Zeitz had an excellent library and, being an educated man, had corresponded with Leibniz on a regular basis since 1711. Leibniz, himself, enjoyed visiting Zeitz where other friends of his also lived and worked.

One of these was Christoph Enoch Buchta. He had, in 1710, become Court Counsellor at Zeitz. Orffyreus had moved to the village of Draschwitz, only an hour's coach drive from Zeitz. During his time at Zeitz, Buchta was able to provide much information to both Wolff and Leibniz on events surrounding Orffyreus' wheel.

Another of Leibniz' friends at Zeitz was the court cleric, Gottfried Teuber (1656-1731). Teuber graduated from Jena University where he studied Mathematics. As well as being a cleric, Teuber organised an astronomical observatory and later became a member of the Prussian Academy of Sciences in Berlin. Leibniz's mechanical calculating machine was of particular interest to Teuber and he was employed by Leibniz to hire a young mechanic called Has, to work on the device. Leibniz was still researching the Guelf family history but seeking a more rewarding occupation and hoping, on George I's accession to the British throne, to be able to travel with the King to England where he had many more correspondents and friends.

On the 19th January, 1714, Teuber wrote to Leibniz,

'. . . I have some very important news for you. A man of the medical profession, called Orffyreus, has constructed an alleged perpetual motion machine in the nearby village of Draschwitz, to which he recently moved. This machine was shown to Mr. Buchta and I. It is a hollow wheel of wood, ten feet in diameter, and six inches thick. It is covered by thin wooden planks in order to hide the internal mechanism. The axle is also wooden, and extends one foot beyond the wheel. It has three teeth which are for moving three wooden stamps similar to those used in pounding mills. The stamps are quite heavy and are lifted and dropped continuously. The iron journals move in open bearings so as to show that neither deception nor an external energy supply are necessary to the machine's motion.

'Having made an appointment with the inventor, we approached the machine and noticed that it was secured by a cord to the rim of the wheel. Upon the cord being released, the machine began to rotate with great force and noise, maintaining its speed with out increasing or decreasing it for some considerable time. To stop the wheel and retie the cords required tremendous effort. The inventor is asking for one hundred thousand Thalers to reveal the mechanism or sell the machine.

'What does your excellency think about this machine?'

Leibniz's reply on the 21st February, 1714 included the following remarks,

'I cannot believe that someone has invented perpetual motion. In my opinion, it is contrary to nature's laws. I suspect that what you saw in the wheel was the action of highly compressed air. However it would need to be recompressed after a short time.

'If the machine was able to show continuous movement for twenty-four hours whilst under lock and seal; or in the presence of witnesses it could remain in motion for twenty-four hours without any additional input of energy, and if it could still apply great force, for example lifting heavy stamps, it might be very interesting. But even then the motion might not be purely mechanical, but rather, physical.'

After posting this letter, Leibniz must have pondered on the Orffyrean machine further. On the 28th March, 1714 he wrote another letter to Teuber completely devoted to the machine.

'Previously I wrote to you and Mr Buchta that I do not believe that mechanical perpetual motion is possible, however it seems to me that if a machine has been made which could remain in motion for twenty-four hours and *still* show a force comparable to that of several men or a couple of horses, then it could still be of great importance — even if it needed its power restored after twenty-four hours. So if you saw such a machine in the village of Draschwitz, then it is of great value. I think it would be beneficial if several trustworthy men could witness and testify that the machine was observed in continuous operation for twenty-four hours.'

Leibniz had realized that a proper test was vital to ascertain the value of the machine, so when the next two letters arrived, one each from Teuber and Buchta, he must have had his already aroused curiosity and interest, heightened. On the 15th April, 1714 Buchta wrote,

'Mr Orffyreus has no objection to running his machine for four weeks, in order to test it! He also asked me to find out if Your Excellency would be willing to buy the machine on behalf of some great prince. He is a quite extraordinary man and is well-worth Your Excellency taking the trouble to see him when Your Excellency visits us. It is only an hour's travel from Zeitz.'

Teuber's letter of the 26th April, 1714 was equally enthusiastic.

> '. . . . I come now to the inventor of the perpetual motion machine and the test of its movement. The inventor assures us that it could run, not just for twenty-four hours, but for a whole month! The machine is utterly amazing! Of course, I cannot decide what principle lies behind it, but I am certain that it is not moved by compressed air contained within it, because there are many slits in the wheel. The increase in its power depends on an increase in its diameter. The model which I saw and which can still be seen in the village of Draschwitz is ten feet in diameter and it has the power to lift about one hundred pounds.'

Leibniz' next letter advises caution, and he is still convinced that compressed air lies at the heart of Orffyreus' invention. His letter, dated 26th May, 1714 says,

> 'Thank you very much for all that you have told me about the machine at Draschwitz. To make sure that everything that you have told me is correct, you should consider the following. Open slits do not necessarily mean that compressed air is not inside, because the external surface need not form the container into which the air has been compressed. Compressed air alone would be insufficient; there must be some other means of applying force to restore the machine to its previous state. I would not dare recommend the purchase of the machine as a great invention, to any great prince, before I or some other reliable and experienced person had examined it.

'However, if it is what it claims to be, I think it could be of great importance and worthy of support and reward. I am very pleased to hear that my calculating machine is progressing well.'

Leibniz next letter to Teuber, continues to reflect on the source of power within Orffyreus' machine. On the 21st July, 1714 he muses thus,

If your Orffyreus' machine contains a strong internal force that can last for several hours or even days, I cannot see any other possibility than that compressed air is inside it.I'm sure that the thin wooden covering cannot contain air, but that does not mean that there isn't something under it that can do so.Certainly, mechanical perpetual motion stemming from such a source is impossible; in which case the source of the force must be outside.If one excludes animals, water, wind, remote weights, a servant rotating it from inside, which isn't suitable for lengthy operations — it seems to me that only air remains. However, we will not be sure until it has been examined. If the inventor can fulfil his promises, I'm sure he can expect the sum of money he is asking for.'

Teuber, the cleric at Zeitz wrote to Leibniz on the 30th July, 1714.

'I don't know what to say or think about the Orffyrean machine. The effect is clearly evident but the real cause is hidden. What is so odd about the machine is that it is moving together with all its parts, including the axle. I would support your hypothesis that the only moving agent could be compressed air, but for the fact that an external source necessary for this is absent. The air contained within is, in my opinion, not sufficient to drive the machine. So even though your theory seems to be the only possible one, doubts still remain. In the meantime, many spectators from different places come to see the machine, and some are of high standing. Our Serene Highness (Moritz-Wilhelm, Duke of Zeitz) came to see it recently with his own eyes, and greatly admired it. I discussed it with several interesting and educated people from Prussia, Brandenburg and Saxony. They were all very enthusiastic about the machine. I would be very pleased if Your Excellency could examine it as soon as possible'.

On the 14th August, 1714 Leibniz wrote to Teuber,

> 'I would like to calculate how much force is exerted by the Orffyrean machine. In other words, how heavy a weight can it lift? How high and how many times an hour? And as well, how many days can it run continuously without interruption? It should be rigorously examined in order to exclude any possibility of fraud involving the application of some hidden force. His Serene Highness (the Duke of Zeitz) wrote to me indicating that he intended to make a more careful, thorough examination of it. It's true that the air alone cannot move anything without something from the outside either propelling it or resisting it. As all other possibilities have been excluded, the air can only come from outside the machine. God-willing, I shall leave Vienna before the winter and go to Hanover. I will try to see you on my way so we will be able to discuss everything then.'

Leibniz was certain that when George of Hanover succeeded to the throne of Britain, he, Leibniz, would be required to travel with his King to England. When the news of Queen Anne's death reached him, he headed straight for Hanover from Vienna via Zeitz, hoping to travel with George to England. On his way to Hanover, Leibniz stopped off at Zeitz to renew old friendships and to see Orffyreus' machine in action. He stayed at Zeitz from the 9th to the 12th September, 1714.

Information about this visit has survived in a letter written to Areskine (Erskine), the personal physician to Peter the Great, the Czar of Russia, by Leibniz. The letter wasn't written until the 4th August, 1716, but I am reproducing an extract here, because it tells us of Leibniz's visit to Orffyreus during that September rush to Hanover. Leibniz wrote that,

> 'Orffyreus is a friend of mine, and he allowed me, sometime ago, to carry out some experiments with his machine. It ran continuously for two hours in my presence and demonstrated considerable power. However, I could not remain there to observe it moving any longer because I was travelling in the coach and with a gentleman of the Duke of Zeitz. I advised him to arrange a test in which his machine would be run for several weeks with all possible precautions taken to exclude any suspicion of fraud. At the same time, data could be obtained about the machine's performance and its power. Once this has been accomplished I am sure that several princes could combine their resources, as he requests, in order to pay him a worthy recompense for his invention. Even if this device is not a perpetual motion machine, about which there

> is so much talk at the present time, it would be still be of great use if it could pass this test of several weeks. He has promised me that he will arrange such a test.'

After his visit, Leibniz continued on his way to Hanover where to his immense disappointment, George not only refused to allow him to become Court Historiographer at the English Court, but deferred payment of his salary until the end of the year. This was a rebuke for staying in Vienna without permission. George told Leibniz in no uncertain terms to stay in Hanover and finish the Guelph family history.

Once settled back in Hanover, Leibniz wrote a memorandum on his visit to Orffyreus. One can almost hear him speaking as he jots his thoughts down on paper,

> 'There is something extraordinary about Orffyreus' machine and we must not ignore it, because it could bring tremendous benefits. If it was proved after a thorough examination, that a wheel of larger construction could do useful work, and if it could be used for example, to drain water from a mine, then I think it would be wise to offer the inventor a considerable sum of money. This money could be paid, either in a single payment or over a relatively short period of time. So as to benefit both the public and the inventor, the following considerations must be born in mind:
>
> **1.** A thorough examination must be carried out to ensure that a larger machine will be of benefit. This is to exclude the possibility of anyone laying himself open to a charge of malpractice at Court by recommending an unreliable device.
> **2.** An agreement should be reached with the inventor for a certain sum of money to be paid to him, if possible, and which he will stand firm on.
> **3.** It would be advisable to pay him immediately some of the money as an interim measure for his subsistence, and to pay him a yearly pension until the time comes for him to receive the complete reward'.

On the 23rd September, 1714, Leibniz wrote to Teuber,

> 'I wish Orffyreus could be released from the burden of supporting all of his family and friends, who are little help to him. Through the

> kindness and persuasion of His Serene Highness (Moritz-Wilhelm, Duke of Zeitz) this could be achieved. I hope he will help him, because his invention merits it. I want him to move his machine to the Zeitz territory. Doing this will ensure the safety of this matter for the future. Will you discuss this with Court Councillor Buchta.'

Having prepared the ground, Leibniz wrote to the Duke himself to ask for his assistance in helping Orffyreus. On the 7th October, 1714 he wrote to Moritz-Wilhelm,

> 'I believe that Orffyreus' invention is very important and I would like it to be utilised, the sooner the better. I am thinking of certain applications within the mining industry. At present he lives under the burden of some people who, it appears, are not acting with his benefit in mind. So I think it would be advisable to consider how to free him from that burden. What ever is decided, it would be wise to take steps to ensure that he is not pressured into moving to another area because he is not faring well - somewhere where he could more easily be deprived of the fruits of his invention. If your Serene Highness in Your generosity could assure him of a sufficient subsistence for a short while, I would hopefully find, in the meantime, a suitable course of action leading to a deserving reward for Orffyreus and the public employment of his wheel. I am writing to Court Councillor Buchta and the Court Pastor Teuber about this matter also, as they can provide you with the best assistance in this matter.'

Orffyreus mentioned in his *Apologia*, that 'a worthy Saxon Duke, no less, offered me the most beautiful of dwellings, and much gracious patronage to go with it. What a pity that good friends of mine saw fit to use force to prevent me from accepting the offer'. Moritz-Wilhelm was almost certainly the 'Saxon Duke'. The inventor seems to have surrounded himself with a considerable number of 'friends' who, as Leibniz commented, did not seem to have his best interests at heart. There are several hints throughout the inventor's life, that he supported a number of people. They may have been the 'poor' that he is supposed to have given money to, over a number of years. One can imagine that they might have stayed close to their benefactor in order to receive the benefit of his charitable gifts. It follows that if they thought his invention was capable of earning its inventor great riches, then they as recipients of his beneficence, would take a strong interest in keeping control of the inventor's negotiations. They may have suspected Leibniz's motives, attributing to him, similar ones to their own. Certainly rumours did surface which cast aspersions on Leibniz's good intentions.

Orffyreus was forced to leave Draschwitz around the end of November, 1714 and, as he always did, 'smashed his wheel to pieces'. It is worth pointing out that every time he moved to a new location, he is reported as 'smashing his wheel'. The words are doubtless Orffyreus' as he always uses them, and reports from newspapers and letters also use them. The emotive quality of the words disguises the fact that the safest way to transport the wheel without risking the danger of someone attempting to steal it, and thus the secret, was to disassemble it. It was merely a security precaution. Even when he died, the one remaining model of his machine was found in pieces, and it made sense to take such action to protect his invention. What it also does show is the very emotional nature of the man. He always blamed his moves on the actions of his 'enemies ' and 'friends ' — their behaviour 'forced ' him to change address, and as part of the blame he would attach the act of 'smashing ' the wheel.

He wintered in Obergreisslau, and then moved again in the Spring, to the village of Merseburg, about eighteen miles from Leipzig.

His stay in Obergreisslau did serve one important service. It gave him time to think. One of the compelling reasons for his lack of success in selling his idea initially, was the fact that there was no proof that wheel was not wound up. The first machine was very small and open to that criticism. The second one was too large to have been driven by clockwork, but that did not stop the same criticism being offered. Now he was able to ponder on the problem, and to come up with a solution. If his wheel could turn in either direction, at will, that should confound the critics on that point at least. If it had a winding mechanism, then one direction of rotation would unwind it, and the opposite would wind it up. How he achieved this is not known, but we shall return to the subject later, when we investigate what turns out to be an excellent clue as to the motive power of his machine.

Once settled in Merseburg, Orffyreus set to work on his third wheel. In his *Apologia*, the inventor often refers to wheels in the plural. History only records the existence of four wheels — the Gera, Draschwitz, Merseburg and Kassel ones. However I have discovered evidence of three more making a total of seven. It would appear that there was more than one wheel in existence at certain times, and this may explain the fact that some versions of some news reports contain descriptions of wheels with dimensions at variance with the usually accepted measurements. There is a lot of confusion over measurements, particularly in Germany where there were nearly as many variants to the ell as there were towns! However Orffyreus specifically states that one of his wheels (the Kassel one) was six Leipzig ells in diameter, and since he was close to Leipzig, I have taken their ell as the standard. There are 1.858 feet to the Leipzig ell, giving a diameter equivalent to 11.15 feet (3.4 metres) for the Kassel wheel, the biggest wheel that he built.

The Mersburg wheel was the first to be able to rotate in either direction, and the inventor makes no mention of any restraining bolts or cords, and in fact, unlike the first two wheels which started as soon as they were released from their restraints, the second two required a gentle push to set them going.

The Duke of Zeitz wrote to Leibniz on the 17th January, 1715,

> 'Orffyreus is at present living in a village near Weissenfels in a wretched farmhouse. Some say he will not build his machine again, but others say he will build another machine much larger and better.'

Weissenfels, close to Obergreisslau was about half way to Mersburg from Draschwitz. The conflicting information contained in the Duke's letter to Leibniz must have caused him concern, and he wrote to the Duke of Zeitz and to Teuber on the 16th March 1715, asking for more information about Orffyreus. But as we know, Orffyreus was thinking of ways to improve his wheel. On the 14th April, 1715 Teuber wrote to Leibniz,

> 'There is a silence about Orffyreus. I don't know what he is doing or thinking. I am afraid that things have gone wrong for him . . . '.

Despite the lack of information about Orffyreus there was news of a kind. In January of 1715, the *Acta Eruditorum*, a scientific publication devoted to informing men of learning about the latest scientific theories and findings was published. In it was an article by Leibniz's friend and erstwhile pupil, Professor Christian Wolff. Wolff was Professor of Mathematics and Philosophy at Halle University. Halle itself was very close to Leipzig and therefore Mersburg, so educated people from those areas must have been familiar with the Orffyrean machine. An abbreviated version of the article runs thus,

> '. . . A perpetual Motion has been built by Orffyreus, a man skilled in the art of Medicine, from which he derives a living; and in Chemistry and Mechanics in which he is versatile.'

Wolff mentions the attempts at determining Longitude at sea by two Englishmen; and the squaring of the circle by a Dutchman; and then,

> 'Ahead of them all is the achievement of the German; the ingenuity of the English has not advanced the solution much; the demonstration of the Dutchman proved unsuitable. But the perpetual motion which our Orffyreus built, has been seen by thousands of people, including experienced mathematicians and mechanics and they were all full of admiration. The mechanism, for which the inventor expects a sum of money, is carefully hidden, and is said to be simple. The diameter does not exceed five Leipzig ells nor does it exceed six inches in thickness. Over a period of one minute, fifty revolutions were observed and the wheel rotated whilst freely suspended with no apparent external source of power. It can provide impulse to remote mechanisms and easily lift weights of sixty to seventy pounds to a considerable height, in a repetitive, equable and continuous movement. This noble invention has been displayed to the public by the inventor in the village of Draschwitz, not far from the town of Zeitz; but he is considering moving to a new location.'

This wonderful piece of advertising arrived at a time when Orffyreus was *in communicado,* so he was unaware of it for a while. According to the *Apologia,* Orffyreus moved out of Obergreisslau just after Easter, 1715 and settled in Merseburg. The new wheel caused him problems to begin with. It needed several adjustments before he got it working to his satisfaction. During his stay in Obergreisslau, Orffyreus received the first contact by an emissary from Karl, the Landgrave of Hesse-Kassel, who was to give him his first big break by offering him a job as Commercial Councillor of Hesse.

Weissenfells was the home of the Saxe-Weissenfells family, and at the Court was Karl's resident, Nathaniel von Stapff. Von Stapff was at one time, tutor to the Landgrave's youngest sons, and also one of Leibniz's many correspondents. On the 28th March, 1715 von Stapff wrote to Karl thus,

> 'May I submissively request Your Highness to be aware that that mechanic, who is working on the alleged perpetual motion device has let me know that he hopes to have his machine ready soon. The moment this happens I will send you a report.'

Karl must have been aware of the invention for some time. Regular communication with his cousin, Moritz-Wilhelm at Zeitz, would doubtless have kept him up to date with the events surrounding Orffyreus. He wrote back to his resident at Weissenfells,

> ' We are very pleased to hear the good news that the mechanic is hoping to have his famous perpetual motion machine ready soon. We would like to know if he could, for curiosity's sake, bring everything here to show Us it working, and how much would something like this cost? It would give Us great pleasure to have more information about this.'

Two weeks later, on the 30th April, 1715 von Stapff replied from Leipzig. Orffyreus had just left Obergreisslau for Merseburg,

> 'The inventor of the perpetuum mobile has assured me that he will be ready with his machine at the Easter holidays. I have, according to Your Gracious orders, not only contacted him again but have also collected information in Weissenfells. Unfortunately he has left their territory and moved to Merseburg, because he felt those High Circles lacked confidence in him. This is due to the fact that he fell out with his companion, which he was quite frank and open about when I spoke with him. This change has been the reason why he could not construct his machine. In the meantime, he showed me several things, but it seems to me that it cannot be strong enough to rotate such a large wheel and, more important, lift a weight high.
>
> 'On this occasion, I tried to find out, for my own curiosity, whether he might decide to come under the protection of, and in a residence of a great prince where he could have a better opportunity to improve his machine, and with Your Highness's support, receive the desired reward. I found that he was not completely unwilling to do so, but he asked me for more time for reflection, and for the name of the great prince who is willing to do this favour for him. I ventured the name of Your Serene Highness and I noticed that he was pleased to hear it. He promised to inform me about his decision, which I will not fail to report to you.'

While these events were running their course, Leibniz stagnated in Hanover. He was desperate for information on Orffyreus' welfare, and his search for a buyer for his machine. Scarcely a week before the von Stapff letter, Leibniz had written to Teuber,

> 'If you find anything out about Orffyreus, please let me know.'

Another of the old philosopher's contacts at Zeitz was Johann Ludwig

Zollman and on the 22nd May, 1715 Leibniz received a letter from him, from Leipzig. It contained some positive news at last.

> ' . . . I must tell you that our famous Orffyreus, inventor of the alleged perpetual motion machine, being in poor health, has moved close to Merseburg where he has rented a house. There are a number of sceptics here, in particular from Mr. Gärtner's circle in Dresden. Mr Buchta related various items of interest to me concerning Mr Gärtner, after his recent visit to Dresden. According to my information he tried to verify his invention, in a minor way regarding its perpetuity, by locking it and sealing it in the presence of a notary and witnesses, and letting it run for an appreciable length of time.'

Gärtner, as model-maker to the King of Poland, had to produce items of amusement and complexity to amaze the Court from time to time. He is acknowledged as an excellent instrument builder, making ingenious clocks and other toys for the entertainment of the court. It is also common knowledge that he would build apparently perpetual motion machines, and then acknowledge that they were fake. Although their ingenuity is undoubted, their requirements were minimal, in that they were only expected to run for a few hours at a time, and were not required to do work. In the main his devices were novelty clocks and the like although there are descriptions which defy belief. One such account tells of a machine in the form of a canvas covered drum which once set spinning would continue to run for an hour. When other such drums were brought near to the first drum, they too began to spin! This smacks of trickery in just the way the model-maker was accusing Orffyreus. Perhaps he could see his position of pre-eminence at the court under attack, deliberately or otherwise, by Orffyreus.

Following the news from Zollman, Leibniz heard from Buchta, who was back in Zeitz. On the 26th June, 1715 he wrote that,

> 'I've been told that Orffyreus has begun to build a larger version of his machine. He is currently living in Merseburg and is in poor circumstances.'

Leibniz's reaction was one of pessimism. 15th July, 1715,

> 'I see from Mr. Buchta's letter that Orffyreus is building a larger machine. I am afraid that his time and labour will be wasted. In trying to achieve too much he may be deprived of all the fruits of his work.'

In the local paper (Leipziger Post Zeitungen) published for the week of June 24th – June 30th, 1715 there appeared an article concerning Orffyreus which seemed to belie the pessimism of Leibniz. Teuber passed a copy of it to Leibniz for his information.

> 'An illness which has afflicted the famous Mr. Orffyreus has prevented him from exhibiting, at the Eastern Leipzig Fair, as he wished to do, a new and larger version of the true perpetual motion machine, invented by him. It will be of interest to the public to know that the famous inventor has just informed us that he has brought to perfection a new perpetual motion machine. It was finished with God's help, just before the recent Whitsun Holiday, in his rooms in the Green Manor by the sixth Gate.
>
> 'The machine is similar to the one constructed in Draschwitz a year ago, and is in the form of a wheel. However, the present machine is slightly larger, having a diameter of six Leipzig ells, (11.15 feet) and a thickness of half an ell, (11.15 inches). The axle or shaft on which the wheel is fixed averages six inches thickness and is six feet long. The continuous impulse attributed to this very useful and ingenious machine, is independent of any hidden power source or drive. According to the original factual description given by the "Leipziger Gazette" on the 3rd November last year, the rotation is caused, without any evidence of fraud, exclusively by its own internal power and artful design. It moves in a fast, strong and uniform rotation as long as the materials from which it is made, i.e., wood, metal etc. last. A considerable force has to be applied to brake or stop the machine. Having put into motion its own mechanism, the power is sufficient to raise high or pull a load of approximately one hundred weight. To demonstrate its effectiveness, four stampers of considerable weight have now been attached. They are lifted by means of eight cams fixed to the shaft, and at each revolution they are lifted twice. Consider how much more such a perpetuum mobile could do if, in the future, a larger one was constructed and by combining several machines have its power multiplied; or it could be applied to other mathematical inventions . . .
>
> 'This new perpetuum mobile has something unique and extremely valuable about it that was not to be found in the Draschwitz model; it can rotate in either direction as desired. As a result it can, turning to the left or the right, wind a load up or down, in or out as necessary. However, to start it, a preliminary push or help by hand is needed.

'The author and inventor is willing to allow those who are of high rank or those who love art and wisdom to acquaint themselves with the machine. But he requests and reserves the right for himself, with all due modesty, to respect the vow he has made to the Almighty God for the gracious gift of this secret hitherto unknown to the world, and which has been sought in vain until now by so many other mechanics, and in the end thought to be impossible. The vow is that on Sundays and on Holidays that are observed as Sundays, he will abstain from exhibiting his perpetual motion machine. But so as to allay suspicion in the minds of those who might try to discredit Orffyreus' invention because of this reservation, he is willing to accept the following arrangements if distinguished gentlemen wish to observe the continuous movement of the wheel for an entire day, or several days, he will permit this to their satisfaction. If the rotation is to last longer than six days, or until it falls on a Sunday or a Holiday, both the room and the machine inside can be sealed or guarded in order to avoid any suspicion. Prior to this, everything must be subjected to a thorough visual inspection from outside.'

The week before the above article appeared, the inventor had sustained a serious head injury, incurred in a fall. He said it happened in the 'loveliest days of the summer of 1715' and he was put to bed. He was still in bed a month later when, at a very early hour, Gärtner, Borlach and Wagner arrived to view the machine. Due to the early hour, doubtless a deliberate part of their stratagem, and the fact that the inventor was ill in bed, another important factor, the visitors were seen by his assistant, probably his younger brother. The 'enemy', using their superior rank to overawe the young man, pressured him into showing them the wheel. A short examination was all that was required, and then the three men left. Their purpose may have genuinely been to define the authenticity of the machine, but surely if they were sincere a longer examination would have been necessary. Their intention was to have a quick look at the device, in order to satisfy themselves that a picture which they had commissioned, ostensibly showing how the wheel was *really* driven, would concur with the actual outward design of the wheel. They had no interest in learning how the machine worked in reality; they may well have felt certain that he was a fraud. They wished to damage the inventor's reputation, with a cartoon of how they assumed that it was operated, i.e., a servant, secreted behind a wall adjacent to the wheel, pulled on a rope which was connected to a rocking beam, hidden above the wheel, perhaps in the room above. A second rope ran down from the other end of the beam, inside one of the supporting posts of the wheel and connected by some mechanism to the bearing or journal on the end of the wheel axle, and thus turned it. A method, in my opinion, as difficult to believe as that suggested by the maid in her affidavit.

Orffyreus became aware of the reason for their visit, when he came across a scandalous leaflet containing a copper-engraving of the above-mentioned cartoon, and bearing Borlach's name. Orffyreus was so dismissive of this attempt at besmirching his name that he included it contemptuously in a later publication.

How long after Gärtner's visit Orffyreus remained in bed is not known, but we know Gärtner wasted no time in publishing his libellous document. He had seen the machine on the 22nd July, 1715 and yet only ten days later, Teuber reported to Leibniz, on the 1st August, that,

> 'Mr Gärtner from Dresden has been boasting that he has detected fraud in Orffyreus' machine. But how can he have detected it? Did he open the machine? I don't know, but I hesitate to believe it.'

Leibniz responded quickly, writing back to Teuber only two weeks after the above letter was written,

> 'If Mr. Gärtner has already detected Orffyreus' artifice he should be able to imitate it. As far as I am concerned, as I have said often, I do not regard that movement to be solely mechanical but stemming from some physical principle. But what it is I still cannot puzzle out. It will be useful because the machine can exhibit considerable energy for an extended length of time. In which case, I cannot call this work of skill a fraud, if it is able to deliver what is expected of it.
>
> 'In all descriptions of its effects there is always a lack of information about how long and how high the weight is lifted. In fact no one has informed up 'til now;
>
> **1.** What weight is lifted during one revolution?
> **2.** How high such a weight is lifted in one revolution?
> **3.** How many revolutions are performed in one hour?
>
> One could then, at least, estimate the power of the machine.'

One can sense Leibniz's irritation at the lack of precise information about the

Orffyrean machine. The next letter from Teuber, although equally quickly despatched contained further depressing news. On the 26th August, 1715 he wrote,

> 'Once Buchta gets back I will visit Orffyreus to examine his machine. I will try to obtain the information you asked for. Meanwhile there are rumours that some dishonest people approached an itinerant who was secretly incited to convince Mr. Orffyreus that Your Excellency is asking about the wheel's secret in order to copy it. In view of this, he should refrain from revealing his secret to you by a vow of silence. I am sure that he will reject his suspicions as soon as he has seen through the deceit and understood its malignancy. But I will ask him myself about this affair and its veracity.'

Buchta also voiced his opinion of Gärtner's actions in a letter to Leibniz, written on the 27th August, 1715.

> 'I do not know whether Your Excellency has received the Gärtneriana. I have been unable to discover from Mr. Gärtner what he is planning. This excerpt from a letter written to him will tell you of recent events connected with Orffyreus' machine. I hope that because of this we may soon know what to expect of this machine. But I am afraid that Mr. Gärtner is going to lose a good deal of his reputation.'

Orffyreus' friends were sure that Gärtner had gone about things wrongly. If he had been a genuine sceptic, he would surely have awaited the results of a test before making his damaging assertions. He did press strongly for a test of the machine's ability and authenticity, but the demands were couched in the most sarcastic and provocative language, designed, I think, to stir Orffyreus into some precipitate action that might result either in the machine and its inventor being proved a sham; or the real secret being accidentally revealed. Gärtner could then either claim foreknowledge of the secret or make a copy of it to amuse the court. Either way Gärtner hoped to maintain his position at court, at Orffyreus' expense.

Various defamatory publications began to issue from the hands of both Gärtner and Wagner. Meanwhile Leibniz responded bitterly to the news that people were inferring dubious motives from the malignant rumours being spread about him. On October the 5th, 1715 he wrote again to Teuber,

> ' Malicious people are trying to get Mr. Orffyreus to think badly of me, particularly as he is of a suspicious nature. I hear that he talks about me in front of the spectators. He is wrong to succumb to his foolish, petty suspicions and is not worthy of the help I am trying to give him. His friends should dissuade him from this improper behaviour. As far as I can see, by the way he has been conducting his affairs so far, he is not going to achieve much. Even if his wheel is a little larger, it still has only a fraction of the power of a water wheel. I was thinking of some applications, besides timekeeping, which would prevent the loss of the invention, but what can one do with a man who, failing to use his common sense, misjudges other people's intentions. Maybe he should be abandoned to his fate, if he abuses their good will.
>
> 'I do not think Mr. Gärtner will easily discover Mr Orffyreus' secret. I also do not think it is purely perpetual motion, because I believe that to be impossible. If it is, then the machine could be enlarged to give much greater power. But I suspect that there is some physical principle behind it because its power is quite limited.'

After Borlach's offensive cartoon had been published depicting, in his opinion, how the Orffyrean fraud was contrived, the inventor had to act quickly. If he did nothing, popular belief in his machine would be quickly eroded to the point where no potential buyer would accept anything but a thorough examination of the machine's most innermost workings before parting with any money. Orffyreus could not accept such a deal as this; it was too risky, he might lose the secret and gain nothing and all his life's work would have been to no avail. In addition within a few days of Gärtner's brief early morning examination of the wheel, there is Teuber's letter telling of Gärtner's boasting that he has detected fraud in Orffyreus' wheel. According to his *Apologia*, the inventor had received a letter from Gärtner while he was still in Draschwitz, asking 'insulting' questions — and we know his reply was less than diplomatic — so there was no love lost between these two men. There was really only one option open to Orffyreus — he must arrange a much more rigorous and thorough test and examination of the wheel. One that would, once and for all, answer all the questions and call a halt to the insidious rumours of deception.

Orffyreus approached Moritz-Wilhelm, the Duke of Zeitz, with a request to sponsor an official test of his machine.

Chapter Five

THE MERSEBURG TEST

The Duke acquiesced to Orffyreus' plea, and on the 19th October, 1715 an article appeared in the *Leipziger Post Zeitungen* to the effect that the official test of the machine would take place on the 31st October, 1715. It explained that,

> 'Mr. Orffyreus, the inventor of the Perpetuum Mobile, which is in Merseburg, wishes to remind us of the promise he made in an earlier edition *Leipziger Zeitungens-Blat.* He refers to the translocation of the above-mentioned machine to another place and another support. He has found it particularly important to fulfil his promise because of the arguments that have appeared in the press recently. They have been expressed in mocking and offensive terms, but they call urgently for such a test. Also certain people have constructed self-moving machines that appear superficially similar to his machine and yet they do not deny that their's is false. Mr Orffyreus has recently recovered from sickness, and is trying to finish, as soon as possible, a Perpetuum Mobile that he has begun to build. He is absolutely committed, with God's help, to show by sincere demonstration the reality of his invention, in the presence of certain persons and skilled craftsmen. This test is scheduled to take place on the 31st of this month. There will be some renowned Mathematicians and Mechanicians present. At the same time, the inventor expects to receive an impartial opinion, having persuaded the Commission of the truth of his case. Because of this, he regards it unnecessary to respond to those people who have until now, expressed their opposition and doubts to the press.'

Orffyreus, very reasonably, takes this opportunity to tell Gärtner that he won't be writing to him again. He must accept the results of the official test. Again we see evidence of Orffyreus' innocence. Is it conceivable — if he was a fraud — that he would advertise an official test *prior* to that test being carried out, unless he was absolutely certain of the findings of that test? He can only have been certain of the outcome for one of two reasons. The first being that his machine *was* genuine and therefore would pass all tests and no fraud would be discovered because there was no intent to deceive. The second possibility depends on him somehow coercing everyone from the Duke of Zeitz downwards to take part in a massive fraud for no discernible advantage.

On the 31st October, the officials selected to carry out the test gathered in Merseburg. The Duke of Zeitz, Moritz-Wilhelm, had taken care to choose men who would be cross-representative of the various interests. Two Commissioners were in charge of the events, and the members of the Commission consisted of people who were judged to be competent to decide if they were being deceived and who had a particular scientific ability. Others were noted for their practical skills, and the remainder for their trustworthiness and incorruptibility.

The two Commissioners appointed by Moritz-Wilhelm were Julius Bernhard von Rohr (1688-1742), Assessor of the Merseburg Court and chairman of the Commission — and Caspar Johann Bretnuetz. Von Rohr was born in Curland, and graduated from Leipzig University, where he studied Law, Mathematics, Physics, Chemistry and economics. He went on to Halle University where he continued to study mathematics with Professor Christian Wolff. After spending a short time in Holland he returned to Saxony and in 1714, was appointed a member of the Merseburg administration. In 1715, at the time of the Commission, von Rohr was only twenty-seven.

To make a record of everything that occurred at the test, the District Magistrate, Johann Andreas Weise was ordered to attend. His account, plus certificates issued by the chairmen, jointly on behalf of the Duke, plus an additional one by von Rohr alone, adds considerably to our knowledge of the events which took place that day. In confirmation of the unanimity of all who were there, many others who witnessed the test, signed their names at the foot of the certificates in approbation of the findings.

The first Certificate said that following the notice in the press, arrangements were made to carry out the test as requested by Orffyreus. The test was to specifically include a translocation of the wheel from one set of bearings to another, thus obviating the accusation that the machine was driven through the bearings. It continued thus,

'The inventor first put in motion his six ells (11.15 feet) in diameter and one foot thick machine, which was still resting on the same wooden support upon which it had previously been mounted. It was stopped and re-started, turned both left and right as many times as was requested by the members of the Commission, or the spectators. The machine was started by a very light push with just two fingers and accelerated as soon as just one of the weights, hidden inside, began to fall. Within about one revolution, the machine had acquired a strong and even rotation, even when a box was lifted, which had been filled with six whole bricks weighing together about seventy pounds. The weight was lifted by means of a rope conducted through a window by means of a pulley, eight ells (14.8 feet) upwards to the roof, and several Clafter (a Clafter is equal to about 8.2 feet) down into the yard. The box was lifted as many times as was requested.

'Furthermore the Inventor, Orffyreus, in front of everybody, lifted the perpetual motion machine described above, from its original wooden support. The timber posts were carefully examined both from the top and bottom as well as in the middle, particularly where a small cut was noticed. The same careful examination was devoted to the iron journals of the shaft and to the bearings. However there was not the least vestige of suspicion of fraud observed.

'By way of additional proof about its internal motive power, the perpetual motion machine was translocated to another support in such a way that the whole assembly could see over and under, and both sides of the machine; and all present were invited to visit the bearings, but no holes were found, all present examined them with their eyes, but no sign of fraud was there seen. It was possible to translocate the machine and to turn it left and right as many times as was asked by the assembled respectable Commission. The machine regained its strong, even and fast rotation each time. The movement was accompanied by quite a loud noise caused by the internal mechanism which lasted until the machine was brought to a forced stop. Nothing suspicious was found.

'Finally, it should be noted that right at the start, before the machine was subjected to any testing, all rooms above, below and on either side were examined by the Commission. It was also verified that the stamps were not hollow, and no indication of any mechanism moved by a cord was found.

Gründlicher Bericht/

Von dem

Durch den anitzo zu Merseburg sich befindenden

MATHEMATICUM

Herrn ORFFYREUM

Glücklich inventirten

PERPETUO

ac per se

MOBILI,

Nebst dessen accurater

Abbildung/

Wie solches seit dem Monath Junio dieses 1715ten Jahres zu gedachten Merseburg von einer grossen Menge Hoher Standes-Personen, gelehrter Leute, Künstler und Curiosorum in Augenschein genommen, und genau examiniret, auch allda noch zu sehen ist.

Cum Censura & Approbatione.

Gedruckt Mens. Decembr. 1715.

und zu haben

Bey JOHANN THEOD. BOETIO zu Leipzig.

Grundlicher Bericht

The above is a facsimile of the title page of Orffyreus' first publication, which extols the virtues of the Meresburg Wheel.

'All that has been written above is the truth, and has been acknowledged by signatures in our own hand without any reservations, This certificate was issued at the respectful, obedient and proper request of the Inventor.

'signed at Merseburg, 31st October, in the year 1715.

The list of those who actually signed the document numbers twelve, but the indications are that there were very many spectators at the event. Often in these cases the whole court would travel with the Duke to witness some event of great interest, and it is clear that most of the court were there that day. Some of the names are unknown to us now, but those who are familiar, are respected as men of exceptional talent and capability, and it is fair to assume that the others were of a similar calibre.

Three members of the Commission were attached to the court of August II, King of Poland and, as August I, Elector of Saxony. He was a powerful ruler, tough, charismatic and demanding of his subjects. But he was also a patron of the arts who beautified Dresden and presided over the Meissen porcelain factory. This man employed only the most able people, and they were proud to be in his service.

One member of the Commission, employed in the service of Augustus, was Professor Christian Wolff. He was born in 1679, so was only 36 at the time of the tests. He was educated at the Universities of Breslau, Jena and Leipzig and was a pupil of Leibniz, who recommended him for the appointment of Professor of Mathematics at Halle University. His application to join the Royal Society in London was accepted as was that of Leibniz. He was later to become scientific adviser to Peter the Great, and helped to found the St. Petersburg Academy of Science in Russia. He eventually became Chancellor of the University of Halle after a period in exile at the University of Marburg, for reasons not unconnected with Orffyreus. He wrote many works on various subjects such as philosophy, theology, psychology, botany and physics. He popularised many of Leibniz's theories and emphasised that every effect must have a cause, and he developed his own philosophy of rationalism and mathematical methodology. He was highly intelligent, rational and practical and he was also the author of the report which appeared in the January edition of the *Acta Eruditorum* of 1715 about Orffyreus and his invention.

The another member serving at the court of Augustus, was D. Johann Burkhard Mencke (1674-1732). At the time of the test he was 41, and had become editor of the *Acta Eruditorum* in his father's footsteps. He was born at Leipzig and, at the age of 20, had become a Master of Philosophy, after which he travelled through Holland, France and England. By the time he was 25 he had been appointed to the professorship of history and had become famous for the quality of his lectures. Two years later he received a doctorate in Law at Halle, followed in 1709, by the nomination to the post of Royal Polish Councillor. He was also historiographer to Augustus II. He became a member of the Academy of Berlin and of the Royal Society of London. He was well regarded both as a person and as a multifaceted scholar.

M. Christoph Semler (1669-1740), the deacon of St. Ulrich's church in Halle was, besides being a minister, a teacher and had studied Mathematics in Leipzig and Jena Universities. He was 46 when he occupied a position on the Commission and had obtained his Master's degree in Halle in 1697 at the age of twenty-eight, after which he received his religious orders. Eleven years later, in 1708 he was appointed to his position as Deacon of St. Ulrichs. For much of his life he studied mathematics and astronomy, but was also concerned with actual conditions and results rather than ideas connected with various crafts, hence the fact that he spent many years in pursuit of the secret of perpetual motion. His undoubted interest in the Orffyrean machine would have necessitated his presence on the Commission, besides which he would have been perhaps, more aware than some, of the difficulties of faking such a device, at the same time he would also know of the possibilities of deceiving his fellow members.

The last member of the Commission was Friederich Hoffman (1660-1742). He was born at Halle. He studied at both Erfurt and Jena Universities, the latter awarding him his Doctor's degree. He was made professor of Medicine in 1693, and in 1694, Frederick III, Elector of Brandenburg, founded the University of Halle and Hoffman was asked to organise the medical school there. He was also a member of the Royal Society.

It is evident that the members of the Commission were chosen for their intellectual qualifications as well as their status within the community. They appeared to have scoured the Universities of Leipzig and Halle as well as various Court circles for the cream of the intelligentsia to add credibility to the test. To make an assessment of the events that were being played out in front of the witnesses, each one needed to, for the moment, put aside any prejudices such as an unbelief in the possibility of perpetual motion. The reason for doing so being, that disbelief in perpetual motion might obscure the fact that the machine might perform, as all present avowed it did, i.e., as if it were a perpetual motion machine — without actually being so. At the same time a healthy scepticism was essential to eliminate any suspicion of fraud. Fraud could only consist of a means of secretly driving the wheel by some artificial sleight of hand. The only connection with the wheel and the rest of the room and its contents was via the bearings and journals at the ends of the axle — that much was obvious to all present. Those were the areas of closest examination. The wooden support posts were minutely scrutinised from all angles, and checked for solidity, as far as was possible. All rooms adjacent to the test room were checked out before hand, and found to be beyond reproach. This test provided the strongest evidence in favour of the inventor.

The other two certificates provide additional details of the examination. The First Certificate, as it was called, was composed by Julius Bernhard von Rohr and Caspar Johann Bretnuetz. It was actually dated the 21st December, 1715 so would appear to have been written afterwards, as an account by the two men, of the test and the reasons for it. It is brief but positive. It tells us that Orffyreus has complained that 'people have been trying to disparage and malign his invention by all means. Amongst other things, a false allegation has been promulgated that its movement is caused and maintained by means of a hidden cord'. A test was arranged at which 'the official scrutineers were shown that the machine could be translocated to a different place several times . . . From external observations it is obviously a genuine example of what is termed a perpetually self-moving wheel. No concealed artifice has been found such as has been asserted, and the accusations are without foundation or reason. To defend himself and his masterpiece, he requested the appointment of some Commissioners and humbly begged us to issue to him the following certificate, which we, the undersigned, granted as we were graciously commanded.'

The document goes on to describe the machine and the tests carried out, and concludes that 'any suspicion of deception or indications that the machine is moved by a hidden cord has been decidedly rejected.' They then signed and sealed the certificate. The remaining certificate was produced and signed on the day of the test, and was entirely the work of the District Magistrate, Johann Weise.

He relates that following the petition from Orffyreus for an official examination of the perpetual motion machine, he was ordered to attend, with persons from the local regional court, the examination of Orffyreus' machine. 'My duty was to observe diligently everything that happened and record it. I was present from the beginning to the end of the examination, and was at the translocation of the perpetual motion machine together with many illustrious persons.

> 'Firstly, the inventor showed us all around and overwhelmingly demonstrated that his perpetual motion machine had no hidden cord as was falsely alleged. The circular machine is about six ells in

> diameter and has a thickness of about one foot. The inventor started it with the merest little effort. As soon as just one of the internal weight s began to fall, the machine started to revolve with such strength that it turned forty or more times a minute, and it could only be stopped with great difficulty.'

An account is given of the turning of the wheel to the left and the right; and the raising of the bricks, and he adds, 'the most extraordinary thing I noticed was that the machine showed the same strength and speed during the lifting and lowering of the load.' Then the translocation is described, including the fact that the inventor requested everyone to examine the bearings on the support posts as well as the journals on the ends of the axle. 'A thorough examination was performed firstly by His Grace's Commission, and then by me, together with the officials from the Regional Office and Court. During this examination, not the smallest cavity or other defect was found. Because of this, everyone was convinced that the impulse must be maintained from within the machine. Then the machine, now on another support, was started again by an equally gentle push, as described above, and again attained the same fast acceleration. All the mathematicians and other intellectually curious people present observed this and were filled with admiration. The entire machine received the highest praise from all, and the inventor was freed of all the false accusations, suspicions and doubts.' Weise then signed his document.

There is no doubting the sincerity of the members of the Commission, or their capability. Buchta and Wolff were both members of the Commission and shortly after the official test, Buchta had words with Orffyreus in an attempt to clear up any misunderstanding between the inventor and Leibniz. On the 3rd November, 1715 he wrote to Leibniz,

> 'Mr. Orffyreus spoke very openly to me about Your Excellency. He is sorry that he did not accept what Your Excellency offered him in my presence. He told me, in confidence, that some malicious people have persuaded him to turn down your generous offer. He and his friends told me that if Your Excellency would like to visit them immediately, they would even let you see and examine the machine as much as it pleased Your Excellency, but under the condition that Your Excellency will swear on oath not to rebuild it for yourself, and will pay three thousand Thalers. They say that for that money Your Excellency will receive nine thousand Thalers, once the machine is sold.

'I was in Merseburg, last Thursday, along with Professor Wolff, Mr. Hoffman from Halle, Mr. Mencke from Leipzig, and Council Assessor for the Court of Merseburg, Mr von Rohr, who was designated by His Serene Highness to examine the machine.

'I saw with my own eyes, Mr. Orffyreus transport his entire machine five or six steps, taking it from one support and putting it on another support. He demonstrated it in full view, and as a result everything that has been written against him by Mr. Gärtner is considered preposterous.'

Leibniz did not take up the offer. He wrote to Buchta on the 3rd December, 1715.

'I am grateful for the description of the examination of Orffyreus' machine. How highly I value his invention, I would love to know the secret, and I would gladly obligate myself to keep the secret and not imitate it. I would even forgo the part which was profit which was offered to me, excluding the portion that I contributed. However I will not pay for the secret as I believe that I will have paid for it by giving assistance in making the thing more useful and more acceptable. I am astonished that Mr. Gärtner who has already seen Orffyreus' machine, should think that Orffyreus pulled, or ordered pulled, a cord to move his machine.

'I was also very pleased to hear that Orffyreus speaks better of me, because I was told that he complained about me and how I wished to obtain his secret, but I believe that he misunderstood me. I am thinking of certain important applications, but I cannot pass judgement without knowing the secret and, consequently what openings might arise for his invention, which does not deserve to remain buried and useless. As the inventor is in straitened circumstances, he deserves to be given, as soon as possible, a worthy reward while he is still alive, so that he can enjoy the fruits of his labour.'

Only three days later, on the 6th December, 1715, Leibniz sent a letter to Teuber,

'No sooner had I written to Councillor Buchta asking him to send me more news of our machine, than I received your letter. He wrote that he was present at the testing of Orffyreus' machine, which is

> very pleasing. He also said that Mr. Wolff from Halle was there, and I have asked him for an account.'

Professor Wolff obliged, with an account of the proceedings on the 19th December, 1715. This letter, one of many between Wolff and Leibniz, seems to have been ignored by the publishers because of its subject matter. Wolff wrote from Halle.

> 'When Orffyreus exhibited the extraordinary machine which he had built, to refute the malicious rumours being spread that it is fraudulent, I was deliberately present. The mechanic, Gärtner, in particular, who is so famous for his many celebrated mechanical inventions, has distributed in public a copper-engraving on which is a slanderous picture showing how Orffyreus' machine was moved by means of a cord from an adjoining room. We have demonstrated that in reality Orffyreus' wheel is far removed from any such deception. The investigation was conducted in the presence of representatives from the Court of the Duke and other guests. When the machine was ready to rotate, all adjacent rooms were opened and the bearings were completely uncovered. To prevent anyone accidentally seeing the internal structure of the machine, he covered it. Whilst he did this, he did not disguise the fact that the mechanism is moved by weights. Several such weights, wrapped in his handkerchief, he let us weigh in our hands to estimate their weight. They were judged to be about four pounds each, and their shape was definitely cylindrical.
>
> 'I conclude, not only from this but also from other circumstantial evidence, that the weights are attached to some moveable or elastic arms on the periphery of the wheel. During rotation, one can clearly hear the weights hitting against the wooden boards. I was able to observe these boards through a slit. They are slightly warped. When he put the wheel onto another support and reinstalled the weights in their previous positions, he pushed down on an iron spring that gave a loud noise as it expanded upwards. I therefore presume that there is no doubt that the wheel is moved by an internal source of power, but we cannot necessarily assume that it is perpetual. Furthermore, the machine may be of little value to the public unless it can be improved. At the moment it can lift a weight of sixty pounds, but to achieve this the pulley had to be reduced more than four times, making the lifting quite slow. The diameter of the wheel

> is about twelve feet, and as well, the bearing was quite thin, about one quarter of an inch and only a sixth of its length was subject to friction.'

Leibniz was grateful for this precise information and he wrote straight back to Wolff on the 23rd December, 1715.

> 'Thank you for all you have related to me about the Orffyrean machine. I must confess, it seems to me to be wonderful. I wish some important Prince would grant the man a worthy reward and buy it. It would be worth it even if it cannot do much at present. Perhaps, once its mechanism is known, it could be improved and used to advantage. Gärtner has been praised for his inventions, I think he should be made aware that he should show more respect for those who are less well-known. He has done himself no good by dishonestly insinuating that Orffyreus is a cheat.'

Besides the newspaper reporting of the results of the test, a booklet was published which advertised the machine and the test it had undergone. This was the publication referred to by Orffyreus earlier. The date of publication was December, 1715. It is a report of the examination of the machine, giving a diagram of the wheel, but actually reveals nothing of interest as far as the secret mechanism is concerned. It shows the rope coiled around the axle of the wheel, lifting the heavy box through the window, via a pulley. As a marketing exercise it is interesting but there are no clues as to how the machine works.

After the diagram and the key to the drawing, there are a few pages of prose. The booklet describes the search for perpetual motion historically, and why it was wanted. It then goes on to describe the work of Orffyreus and the slander and libel he had to undergo before the successful test. It claims that up to one thousand copies of the libellous engraving by Borlach have been distributed, but that the recent test carried out proves the falseness of its insinuations. Other derogatory accusations are described, such as it needed a violent push to start it; and it could not actually lift any weights. It was, says the booklet, for these reasons that the inventor pleaded with the authorities to test his machine. Various names of those who attended the test are given, plus a few who did not sign the certificate, and then an essay on the merits of the wheel is included, written by Christian Benit and Johann Just. Walbaum, both spectators at the event.

The essay suggests uses such as driving a millstone, or being used for cable-making, glass or stone polishing, alloying, laundry and bleaching, in clocks and various pieces of machinery associated with hydraulics. It goes on to reason that the useful output of the machine could be multiplied by increasing the size of the interior workings, or by making wheels of up to 20 ells in diameter. Putting wheels in tandem, it is suggested is another possibility. Particular interest is shown in the possibilities for pumping water, and the mining regions are identified as potential users, both for extending the working seasons as well as for uncovering new previously flooded seams. The list of potential uses which they go on to discuss seem endless, and I won't bore the reader with any more. Suffice to say that the authors of this booklet were extremely enthusiastic for the potential that they could see in this machine if it was developed. The authors, Benit and Walbaum end their essay with the promise that Orffyreus made prior to the test — that if his machine was found to be a fraud, or did not run for ever until the parts wear out, then he, the inventor, would admit to his 'trickery, refund any money already paid, and place himself at the mercy of the legal process'.

Benit and Walbaum went on to found a school for Mechanicians. A place where the theory of mechanics was taught, something on the lines of which Orffyreus himself had plans to do, although without the strong religious ingredient which he intended.

Earlier we discussed the fact that Orffyreus had instituted a collection for the poor. This was done to try to reduce the numbers of people calling to see the machine. The local authority at Merseburg imposed a tax on this money, but the order originated in Dresden, the capital. The tax was imposed in January, 1716 and negotiations for its implementation would have begun towards the end of the previous year. Gärtner was commonly held to be at the source of this action, and it seems likely that he used his position of eminence to persuade some authority to take this action against Orffyreus, by suggesting that the inventor was ostensibly collecting money for the poor, but actually pocketing the proceeds, and perhaps a tax should be imposed to ensure that the authorities got their share of the profits. Gärtner although seldom lending his name to anything obviously slanderous, seems to have been at the heart of all the mischief that came Orffyreus' way, but he had two willing accomplices in Wagner and Borlach.

Following the pressure placed on him by this tax, and the continued lack of success after the recent test, Orffyreus suffered some kind of break-down. He lost faith in his friends, became ill, and grew more depressed. Rumours of his 'madness' spread along with conflicting reports of arrest and imprisonment. The illness lasted for some four months and may have been some form of nervous disorder due to the various pressures placed on him, and his own volatile temperament, in addition he seems to have been, as Leibniz pointed out, 'in straitened circumstances'. After the optimism that the test generated, he sunk in to despair with the renewed attacks by Gärtner — the tax demand, the slanderous tracts, the rumours etc. If he was really collecting money for the poor, it must have been very hurtful to have his motives impugned in this way. 'Smashing' the wheel was probably the only option open to him anyway. To avoid paying the tax he would have to stop the collections. That meant his house would be besieged again by crowds of onlookers; something he could not stand the thought of. To prevent that he destroyed his machine. Once that was done he might as well move on. Orffyreus' opponents must have thought that they had won, and they might well have done so if it had not been for the timely intervention of Karl the Landgrave of Hesse-Kassel.

On the 4th February, 1716 one of the Zeitz Councillors, Johann Zollman, wrote to Leibniz,

> 'As far as the perpetual motion machine is concerned Court Councillor Buchta examined it closely and, I am sure, has already reported to you. The one thing we can now be certain of is that no fraudulent movement from outside is involved in it. But this still leaves the question of what can physics or mathematics say about it? But still, even if it is genuine, I am not sure that there is sufficient power available to provide benefit for other mechanical inventions. I was told at the Leipzig Fair, that the inventor has fallen into a delirium.'

Teuber, writing to Leibniz on the 7th March, 1716 had also heard the gossip,

> 'There are rumours that Mr. Orffyreus has succumbed to delirium some time ago. But is it true? I just do not know.

Prior to the arrival of Teuber's letter in Hanover, Leibniz had written, on the 10th March, 1716,

> 'Someone wrote and told me that Mr. Orffyreus had fallen into a delirium, which I find terribly sad. I would hate it if such a good man was to end up without any reward for his tremendously important and highly admirable work. I am afraid that he or his friends and associates will not bother to find a really significant application for his machine. I would not worry so much if this was to happen under the supervision of somebody who has some fundamental knowledge and is skilled in advanced mechanics. It has been a long time since I have heard from Mr. Buchta, even though I responded to his letters, and wrote as he required. Perhaps he will write soon and tell me what he is doing.'

A few weeks later, still pining for news of the Orffyrean invention, he wrote on the 14th April, 1716 to Teuber,

> 'I am very worried that something may have happened to Mr Orffyreus, and I would be very pleased if you would enquire about this. The superintendent from his place will know best what has happened. Many things have already troubled that good man, and he has not accommodated to the rules of the communal life. Because of this, I am worried that his invention could be easily lost, if it is not disclosed to his friends.'

Leibniz sums up Orffyreus' nature precisely. He is a good man, but he has not settled to a communal life. His nomadic existence could be to blame, it was in his nature to be restless. The old philosopher seems to have a soft spot for the inventor, and offers no criticism for his, in some quarters, unacceptable lifestyle. Reassuring news arrived from Zollman in a letter written on the 19th May, 1716,

> 'I heard from a distinguished gentleman at the Leipzig fair, that Orffyreus has in fact, not fallen into Delirium but rather, acted in such a way that people concluded that he had. To be precise, he was so angry about the truly disgusting excise duty or business tax, levied on his perpetual motion machine, that he destroyed it. There is no doubt that that trick was engineered by certain malicious people, but even in his anger he should not have destroyed his machine but should rather have dismounted and disassembled it.'

The inventor does describe how he sometimes mumbled meaningless words to his visitors, in order to put them off asking the same old questions. Some of them may well have gone away thinking he had succumbed to mental illness. Zollman makes the point that the inventor should not have smashed his machine, but just disassembled it. A valid point, but maybe Orffyreus, who did this prior to each move, was worried that a simple disassembly might leave the secret of the machine's construction too open to re-assembly by anyone who succeeded in taking the complete but disassembled parts away in a robbery, particularly if the design was a simple as we are led to believe.

Sometime in June, 1716 Orffyreus recovered from his long illness. Whether it occurred naturally, or due to his own ministrations is unclear. The recovery might have been given a boost by a letter he received from Karl, the Landgrave of Hesse-Kassel. The first letter invited him to come to Kassel, the second told him that all expenses incurred in the journey would be met. During June Orffyreus made his way to Kassel to discuss arrangements for moving himself and his family there, and on the 3rd July, 1716 Leibniz, writing to Teuber informs him that he is leaving Hanover to travel to Zeitz to sort out the remaining difficulties connected with his calculating machine. He also mentions that he is looking forward to seeing Moritz-Wilhelm if he is at home. The Court diary records the fact that Leibniz arrived during a celebration at the court on the 10th July, 1716 and was treated as the guest of honour. He stayed there for a week, modifying and perfecting his calculator, and called in on Orffyreus on his return journey to Hanover. On the 3rd August, 1716 wrote to Teuber accordingly,

> 'I spoke with Mr. Orffyreus, who is about to move to Kassel. He has high hopes that, with the support and authority of His Serene Highness, many eminent princes will be approached, and he will be able to get a just reward. But to achieve this, he should allow the machine to run for as long as two weeks.'

Leibniz left Zeitz on the 17th June, calling on Orffyreus on the way back, and was told of the inventor's plans to move to Kassel then. We shall learn later, from correspondence between Wolff and Leibniz, on the subject of the Czar of Russia's interest in the wheel, that it was known on the 22nd July, 1716 that Orffyreus intended to move to Kassel, so the inventor must have already been to Kassel and accepted whatever offer Karl made to him, probably in June, and had returned to make arrangements for the move. Leibniz would have been pleased to hear of the move, perhaps hoping that now, at last the perpetual motion machine would be recognised, Orffyreus would be paid, and the invention released to the world. In addition to this he had set in motion the final adjustments to his calculating machine; it was almost as if he was tidying his affairs perhaps prompted by some premonition that time was short. He returned to Hanover and, on the 14th November, 1716 at the respectable age of seventy, Gottfried Wilhelm Leibniz died.

Chapter Six

KARL — PATRON OR PRETENDER?

Karl the Landgrave of Hesse-Kassel has one of the most important parts to play in the legend of Orffyreus' wheel. What sort of man was he? Would he have connived in promoting a fraud of dubious benefit to himself? Was he the sort of man, easily carried away by powerful argument or convincing displays of apparent sincerity? Was there some other reason why a man of his status would have been persuaded to embark on a deception so grand in its scope that it would have to involve the Czar of Russia, the King of England, the King of Prussia and the King of Poland?

In 1716, when Orffyreus and his family arrived at Weissenstein Castle at Hesse-Kassel, Karl was sixty-two. He had had a long and successful career building up his small country's fortunes and its standing in the world had risen substantially due to his energetic participation in the many wars of this period, and to his ready supply of mercenaries available to those States who could afford them.

In the seventeenth century many of the small states had soldiers for hire, but Hesse-Kassel was the first and one of the most successful. Before Karl's time the thirty years war had raged across his territory, soldiers had marched hither and thither, been quartered in its towns, had laid waste to the land during tactical retreats, and at the end of the war, when the armies were disbanded, the countryside was in a poor state. There was, however, one resource available which could provide an income for the state — the army. The European power struggle in the century following Karl's accession to the Landgraviate of Hesse-Kassel, was beneficial for the business of hiring mercenaries. At the end of the thirty years war, although many of the soldiers were allowed home, three companies were retained to guard the palace of the Landgrave, and these were to form a nucleus of the Hessian Guard, the core of the Hessian army. The great powers in Europe, needing large numbers of soldiers and unable to conscript and maintain large enough armies from their own resources, turned to men such as Karl for additional men. Hesse received subsidies for fighting, from which ever power hired them. Karl's original three companies had, in six years, grown to eighteen companies of foot soldiers and five of horse.

Karl was the second son of Landgrave Wilhelm VI, and was born on the 3rd August, 1654 in Kassel. The early death of his older brother in 1670 in Paris should have enabled Karl to become Landgrave, but due to his being only sixteen, his mother continued to run the small state, even after his eighteenth birthday. She was a powerful and determined woman and when, in 1673, she ordered Karl, still only nineteen, to marry his dead older brother's widow, Marie Amalie, he obliged. Eventually she abdicated in his favour in 1677 and at a grand ceremony he assumed control of the government of his country.

His primary concern was to create an army for himself which would always be ready for war, and he made many sacrifices for it during his years as ruler. He spent a fortune equipping his troops with the best of everything. He fought in several wars, acting as both ruler and general and always led from the front. But in 1699 he took a break, and on the 5th December started a journey to Italy, under the name of Imperial Count von Solms, travelling as far as Naples and taking every opportunity to visit as many treasures and sights of Italy as possible. He left the care of Hesse-Kassel to his son Frederick, later to become King of Sweden. This trip suggests that beneath the tough, disciplined soldier, lay another, more thoughtful man, interested in the arts and sciences of both the past and the present.

Karl concluded subsidy treaties with Holland and England and in 1726 joined the Hanoverian Alliance. In 1730, aged seventy-six Karl died. He was known to be fiercely patriotic and he and his well-equipped troops appeared time and time again as the first in the field of battle, but one thing even more important to him was the protection of Protestantism, allied to improving the wealth of his country. So, to bring new capital and labour into his country, he granted the Huguenots, who had been expelled by the edict of Nantes, shelter in Hesse under the most favourable of conditions. In 1710, he established a Chamber of Commerce which was to promote and supervise the province's products and produce. Orffyreus was made Commercial Councillor when he joined Karl at Hesse-Kassel in 1716 and was part of that organisation.

Earlier, he had organised the construction of a port on the river Weser at a place formerly called Sieburg but later named Karlshafen in honour of Karl, and this small town was to become the place of Orffyreus' demise. Karl planned to connect Karlshafen with Kassel by means of a canal, which was to be extended to the river Lahn, however although considerable progress was made, the complete enterprise was not finished before Karl's death.

Karl's greatest visible memorial, part of his tremendous building activities, was the construction of the Hercules Octagon with its cascades at the Wilhelmshohe, Kassel. The water tumbling down the cascade was pumped to the top of the cascade by the first steam pump to arrive on mainland Europe, an event of vital significance to the Orffyrean saga. Karl's interests included not only objet d'arts but curiosities of all kinds. This included his large collection of antique and modern gems and coins, maps, physical and astronomical instruments which he had brought back from his Italian journey and exhibited in an art gallery built by him.

But it was due to his strong predilection for mathematical and physical investigations and experiments that he acquired the nick-name of a 'curious gentleman', a term in those days used to describe someone with an outstanding curiosity to fathom the secrets of nature. In 1688 he called a Huguenot, Denis Papin, to Marburg as Professor of Mathematics and managed to keep him in his services for almost twenty years. After moving to Kassel in 1695, Papin carried out a series of the most famous experiments which were chiefly concerned with the practical application of steam energy, under the eyes of the Landgrave and with his complete support.

One of the most enduring memories of this man, that has come down to us and will become evident in some of the letters about him that we shall be looking at, was his great affability. But that does not mean that he was a pushover, as those who were involved in the delicate negotiations that resulted in the Hanoverian alliance, will testify to. Envoys from England, Russia, Sweden and Prussia all speak of Karl in respectful terms and often comment on the disparity between the power that this man wielded at the conferences and the actual size of his country. His influence was undeniable. He spent thirty-eight years in happy marriage to his brother's widow, Marie Amalie and she bore him ten sons and five daughters. His final years, after her death, were spent with the Countess Bernhold and the Marquise de Langallerie.

The Marquise was the wife of the notorious Marquis de Langallerie, Philippe de Gentils, a French adventurer and soldier. He had started his career as a soldier, distinguishing himself with his brilliant deeds, fighting in twenty-two campaigns and becoming in 1704 Lieutenant-General. He has been described as proud of character, ambitious and haughty, but he had a habit of claiming for himself the successes of anything in which he had taken part. He died in captivity in Vienna on the 20th June, 1717 of boredom so it is said. Little mention is made of his wife and she, it seems was left largely to her own devices. Langallerie was a Catholic, and yet on the 13th May, 1714 a letter from Paris informs us that 'the Marquis de Langallerie is gone to Amsterdam, and makes profession of the Protestant religion'. This may have made it difficult for him to return to France and perhaps he and his wife were offered refuge by Karl. In any event he left to go on his travels and ended up a captive in Vienna; she spent the remainder of her days at the court at Kassel. We shall return to the subject of the Marquise de Langallerie and her baleful influence over Karl later.

Denis Papin, long-time collaborator with Karl on the development of the steam engine, had been invited to England where his steam experiments had provoked great interest. He was a member of the Royal Society, and had previously been to England where he had worked for Robert Boyle, assisting him in his experiments with the air-pump. In 1679, Papin had demonstrated to the Royal Society, his invention called a 'Digester' which was an early form of pressure cooker. His second visit to England when he left the employ of Karl, was to try to interest the Royal Society in his steam-navigation projects, and to persuade them to compare, through tests, his steam engine and that of Captain Savery, the forerunner of the famous Newcomen, or Beam engine. On the 11th February, 1708, Papin submitted to the Royal Society further proposals for the construction of a steam boat, having paddles moved by fire and steam. This did not materialize, and Papin's quality and quantity of work declined and he, receiving small payments from the Royal Society, eventually disappeared with no record of either his time or place of death.

If Papin had stayed in Kassel, he might have fared better, but his most productive time was over. He was sixty years of age, news was arriving from England of a machine which was to eclipse his work totally, although much of it may have resulted from the early experiments that he had carried out with Karl. Papin sought recognition from the most influential scientific body around—the Royal Society. But another Huguenot, Jean Desaguliers, curator of experiments at the Royal Society, had the ear of the President, Sir Isaac Newton. Desaguliers had his own reasons for discouraging interest in Papin's experiments; he was working with Captain Savery on the new Savery-Newcomen steam engine, trying to improve it. We shall be returning to them later, to see how Desaguliers thwarted attempts by two of Europe's leading scientists to interest Newton in Orffyreus' wheel. Karl, at a loss for a new direction in experimental scientific research to take an interest in, seized upon the Orffyrean wheel, as a possible successor to the Papin experiments.

After agreeing the details of the arrangements for Orffyreus' transfer to Kassel and appointment to the position of Commercial Councillor, Karl allowed his new protégé the use of the apartments at his castle Weissenstein at Kassel. As Commercial Councillor, Orffyreus would be paid a yearly salary of three hundred Thalers (about £60) plus food allowance, and in addition to this he was reimbursed for his expenses incurred moving from Merseburg to

The mighty cascade at the Castle at Kassel, designed and built by Karl, the Landgrave of Hesse-Kassel. Water was pumped from the foot of the cascade to the top by Europe's first steam engine.

Kassel. Mention is made in some letters that will be quoted from later, that Karl had seen the inside of the wheel, and it was evidently part of the deal worked out between them that, for a price, Orffyreus would allow the Landgrave into his secret. Karl was the only man ever to share this knowledge, and it is a compliment to the character of the man that in the first case, Orffyreus trusted him; and in the second, that the Landgrave never broke his promise to keep secret what he had learned. He would not have afforded the inventor his patronage without a guarantee that the machine was not a deception. And he would have insisted that the only proof of Orffyreus' sincerity that was acceptable was the revelation of the secret. It must have been difficult for the inventor to decide whether to take Karl into his confidence, and risk the secret getting out — or to refuse the wonderful offer of a job, patronage and board and lodging, and assistance in selling his machine. Orffyreus was of a naturally suspicious nature. He had lived on his wits for too long to risk losing sole ownership of his secret. Of course the inventor did not give in without a fight, and Karl agreed to pay a price for the secret, four thousand Thalers!

According to Leibniz, writing to a colleague in Kassel,

> 'His Serene Highness, the Landgrave has, not only an intense intellectual curiosity but, more importantly, he has considerable scientific insight and knowledge. It is precisely these qualities that allow the arts to flourish and attracts people of talent. As you know, I am interested in wonderful discoveries, so I am asking you to let me know, from time to time, what His Serene Highness is doing for the advancement of the arts and sciences, and for the benefit and enhancement of His country.'

In another letter to Haas, the librarian at Kassel, Leibniz remarks that,

> 'There are few, in my opinion, who possess the genius of His Serene Highness, the Landgrave. When such a prince exists, one cannot wish him enough life and prosperity because His glorious influence extends not only over His own people but also spreads far and wide, so that all mankind can profit from the inventions that have been achieved under His protection.'

Once established in the castle of Weissenstein at Kassel, Orffyreus began work on his most magnificent creation — the biggest and most powerful perpetual motion machine yet. What with moving his family and effects to Kassel and settling in; coming to terms with his new position as Commercial Councillor; acquiring the materials to build his new machine and, of course writing his biography in rhyme, it was nearly a year before the new wheel was ready, but by August 1717, the machine was being exhibited again.

During that year of 1717, Gärtner, Wagner and the other of Orffyreus' antagonists published more sneering pamphlets, including a repeat of the wager so scathingly mentioned in *Apologia.* This time the publication not only renewed the bet of one thousand Thalers, but made new demands of the inventor. The article appeared in print before the new machine at Weissenstein castle had been finished, and perhaps the authors hoped to forestall the building of the wheel or at least prevent its completion. The Gärtner gang may well have thought that they had seen and heard the last of Orffyreus, so when they received news of his imminent departure to Kassel as a guest of the Landgrave, they swung into action again with a rehash of their previous challenge.

It is long-winded, as was much that was published at the time, (Orffyreus had no monopoly on excess verbiage). In short, the Gärtner 'gang of three' renew their offer to give Orffyreus one thousand Thalers if he can prove that his wheel can turn continuously for four weeks. They remind us that the inventor had asserted that his machine could do this, and if it was not true then the authorities could have his head. They enumerate various valid points, such as that the inventor had already offered to accomplish this if his expenses could be met. Secondly they claim that the inventor sent an offensive letter to the 'Royal Model-Master' (i.e. Gärtner) claiming that he had turned down a similar bet previously agreed to, which would only have cost 100 ducats, which was about 300 Thalers or £60. They claimed that this was a lie and that they had accepted the bet. Personally, I do not believe that Orffyreus would have made a bet of such a low amount, it would not have been worth the trouble to arrange for the four week continuous observation, and tends to throw the rest of Gärtner's claims into doubt. The 'gang of three' go on to suggest that Orffyreus does not possess a perpetual motion machine; that it cannot run continuously for four weeks; and that it cannot lift a weight of seventy pounds.

To assist the inventor in his attempt, they generously allow him to use any methods which he thinks may be of use in his attempt. They include a list of materials, philosophies and strange beliefs too long to rehearse. He can include everything in the list — BUT he must comply with the conditions summarized as follows:

1. All of the things mentioned previously, if used, should be hidden within the wheel.
2. The wheel's bearings and journals should be uncovered and open to inspection.
3.The wheel must rotate continuously for four weeks at fifty revolutions a minute, and be able to lift 70 lbs.
4. Finally, the wheel should not be touched by anybody during the four weeks, under the excuse of making necessary adjustments, unless something should break. In that circumstance, one breakdown will be allowed, and the four weeks will begin again.

Gärtner continues by suggesting that as three of these conditions were taken from Orffyreus' own ideas, and the fourth is a reasonable addition, he need have no fear of getting involved with the test. Furthermore, by this test the inventor will prove that his machine is genuine and he will win for himself one thousand Thalers (£200). No excuses due to time-wasting or money or wear and tear on his machine can be offered as all those points are answered. The test can take place where ever Orffyreus wishes, but it should not be too far from Dresden. Finally, Gärtner refers to the inventor's declaration that he does not gamble. If he wishes to look on the test not so much as a wager, but more as a reward for completing the test successfully, then he can do so. Alternatively he could give the money to the poor.

There is nothing in these suggestion as to how the test should be run, to complain about. All of Gärtner's words are eminently reasonable and the wager is couched in the most moderate terms — so why was the wager not taken up? Firstly one must remember that Orffyreus hated the 'gang of three' with some justification. They had slandered and libelled him in a way no modern justice system would allow. They had tried to trick him into revealing his secret. They had engineered a tax demand on his charitable collections and forced him to end those and to disassemble his machine — and they had released a torrent of malicious rumours with the intent to destroy any remaining vestiges of his good reputation. Add to that the probability that the wager that Orffyreus is supposed to have accepted of a mere 300 Thalers, was a lie for the benefit of the readers of the tract, and one can understand the inventor's reluctance to enter into this bet. He must have been strongly suspicious of the conditions that would actually obtain at the test. He may have feared losing control of the test, with all of Gärtner's cronies present. An attempt could have been made to pry the secret from him directly or indirectly. He may well have suspected that foul play would be used to nullify the test. There were a number of ways that the machine could be brought to a stop; by applying force to knock the wheel off its bearings, or by damaging the bearings or applying too great a weight for it to lift. The stipulation was that only one stoppage would be allowed for repair, and no touching of the machine would be permitted. It is easy to see that, since Orffyreus could not be present twenty-four hours a day, at some stage, perhaps when one of his aides was present, a moment's diversion could allow one of the above ruses could be applied. Two stops of the wheel, and Orffyreus was one thousand Thalers out of pocket.

It was essential that all tests were carried out by people who were at least prepared to view them without prejudice, something one could hardly claim for Gärtner and his friends. The Duke of Zeitz' test had adequately confirmed that the wheel could perform as required by Gärtner. Fortunately, within a month of the above publication, Orffyreus had completed his wheel. The suggestion by Gärtner that Orffyreus would not gamble, is doubtful. The inventor freely admits that he gambled in his youth, and, indeed, offers his own wager in the next document which we shall be looking at.

The appearance of the wager at this point may have been to try to stifle the emergence of the new machine. By publishing the document when they did, they also made it appear as though they had succeeded in forcing the inventor to react to the wager, and announce a test, rather than him arranging the test according to a preconceived plan. As soon as his machine was ready Orffyreus requested the Landgrave to arrange a long-duration test, and this was carried out from the 12th November, 1717 until the 4th January, 1718. That is, fifty-four days in all, almost eight weeks, and far in excess of Gärtner's four week demand.

After the successful completion of the test, Orffyreus published a pamphlet extolling the virtues of his machine. He acknowledges the kindness of the Landgrave in arranging the test and says that it was carried out in a room specifically assigned to the inventor. As soon as the machine had been started, the room was locked and sealed. Guards were placed outside the building. The machine performed perfectly and Orffyreus therefore wished to lay a wager of 10,000 Thalers with 'all those people who, doubting the possibility of a perpetual motion machine, made bets, more than once of 1000 Thalers'. This is clearly a dig at Gärtner. He goes on to recount the actions of the 'enemy' who 'have expressed their fury in satirical copper-engravings, sarcastic pamphlets and malicious defamation'. 'What,' he asks, 'will they say now, since Orffyreus who they thought had been totally discredited, has appeared with a newer and larger perpetual motion machine? And this under the auspices of a great prince, who is famous throughout the world for his wide education and deep understanding of mechanics. The test lasted for one month longer than everyone expected, and was very well executed'. The inventor included a copy of the certificate granted to him by Karl. Also provided was a diagram of the new machine, with a brief description of the parts that were shown, however as before, there were no clues as to what was inside. The description includes the information that the wheel 'is almost identical to the Merseburg machine, because that machine too, has had to be built in a room, so it was impossible to have a larger diameter than twelve feet'. It was pointed out however, that the new machine was six inches thicker than the Merseburg one, being one and a half feet in thickness.

One thing that seems to have been ignored in all accounts of the Orffyrean machine is the inclusion in the above-mentioned diagram, of a pair of large counter-weights on the end of pendulums which are connected to the journals on the ends of the axle. The inventor says that they are designed to slow rotation and give it more uniformity. But, he says, these can be dispensed with. In the accounts we have already seen of the action of the wheel, more than one commentator has remarked on the great evenness of the rotation, and yet no-where is there any reference to the two large pendulums mentioned by Orffyreus. I believe that the inventor may have thought his machine might run wild if no means of governing its speed was incorporated, hence the pendulums, but in fact no such tendency was apparent for reasons which we shall discuss in the chapter on how the machine actually worked, so they were removed.

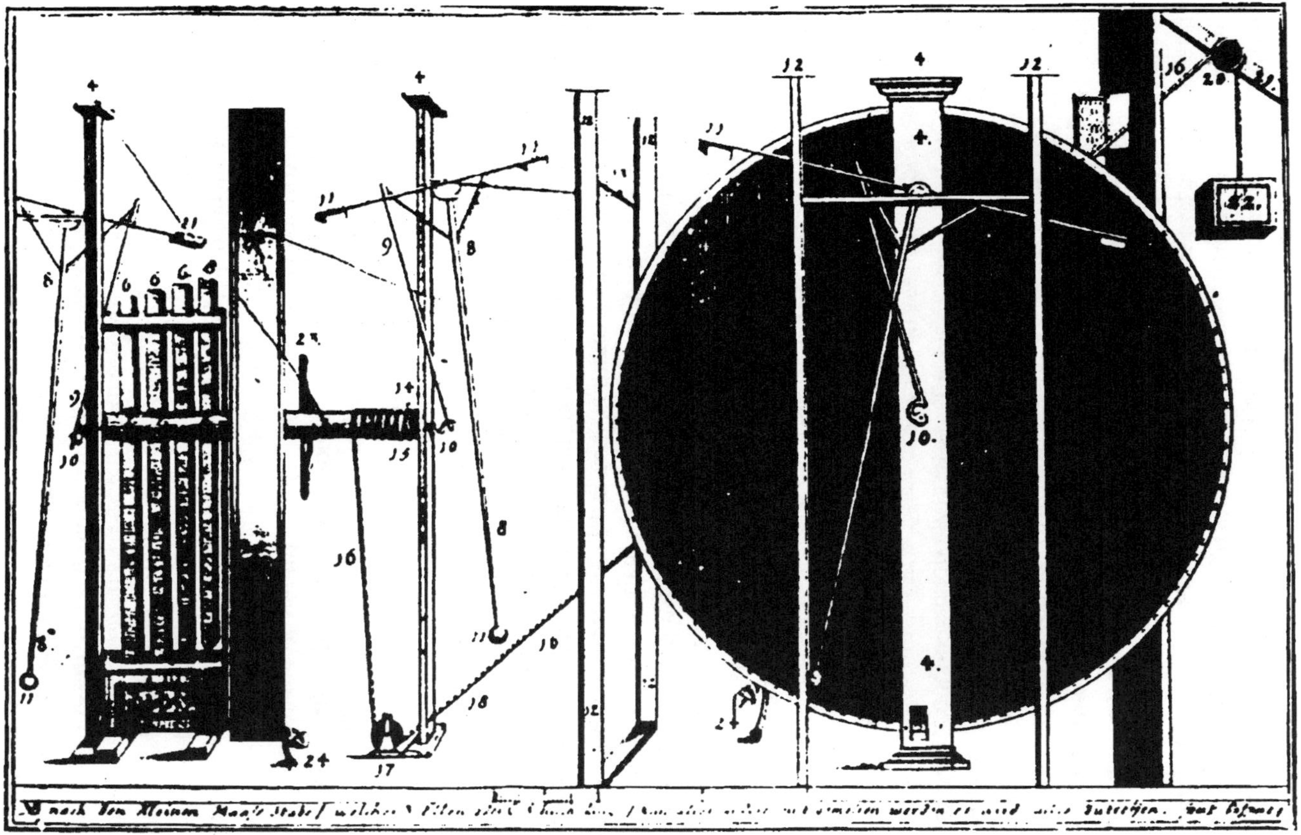

This illustration is from Johann Bessler's first publication, in 1715 – 'Grundlicher Bericht . . .'

Orffyreus finishes by reiterating that his machine does not require winding or the assistance of such things as the power of the wind for its movement. He repeats the offer of a wager of ten thousand Thalers and says that provided his wheel is allowed to rotate according to its internal disposition, in other words at what ever speed it is designed to do, he will allow it to run for any length of time suggested, subject to the materials of which it is made, surviving intact.

He asks in a final mocking challenge, 'Where and when will Mr Christian Wagner construct his "turnspit" which appeared in a copper-engraving some time ago? It is supposed to run exactly like the Orffyrean machine. Similarly, where is the perpetual motion machine solemnly promised to the curious public by Mr. Mahn, Court Cake-Master a year ago? 'These items have been long and anxiously awaited, and they have contributed much to the delay in the acceptance of the Orffyrean Perpetual Motion.'

Jacob Andreas Mahn had been reported in a local newspaper in January, 1717, as having worked on the theory of perpetual motion for twenty years. He claimed to have discovered the secret several years ago, and suggested that it was the same as Orffyreus' machine. He had claimed that his Court duties had prevented him from finishing his machine, but that because he did not want his toil and trouble to be in vain, he had now nearly finished it, and if people would wait before entering into negotiations with Orffyreus for four months, he would be able to finish it. The potential buyers would then be able to buy his machine at a fraction of the price that Orffyreus was demanding. He makes the important point that buyers should beware of obtaining the Orffyrean machine, because his own machine would soon be available and then they would not own exclusive rights to the secret.

Obviously this was a delaying tactic, and it is not hard to guess where it originated. Gärtner was Master Model-Maker to the King of Poland; Mahn was Master Cake-Maker to the same King. They were both members of the same Court, I am sure that some kind of collusion was at work here to obstruct Orffyreus further. Gärtner held a high position at the Court, and we have seen that he was not above using his influence to achieve his aim, he might easily have sought out Mahn and put a proposal to him, to be rewarded with some suitable inducement that it was in his power to give. Mahn admits that he has spent many years searching for the secret of perpetual motion, and we already know that Gärtner had flirted with the idea, so it is almost certain that they had discussed the theory on several occasions in the past. It was a good plan, bearing in mind that any potential buyers of Orffyreus' machine would hesitate to complete their negotiations knowing that another version of the same machine might soon be available, and at a cheaper price. Gärtner, could not allow himself to be associated with such a plan openly — he was, after all a declared opponent to the possibility of perpetual motion.

The certificate that Karl gave to Orffyreus was similar to the previous ones; it described the precautions taken against fraud; it described the tests that the machine underwent. As previously, it recounts the building and successful testing of machines in 1712, 1714 and 1715 in the various towns. It repeats Orffyreus' complaints about the harassment he has received, the libel and slander, and the sarcastic wagers he has been challenged to enter. It is as a result of all this activity, the certificate says, that Orffyreus has requested from Karl the following action.

> '. . . to subject said machine to Our noble gaze in person and, when it has commenced revolving, to watch carefully over it and, in order to forestall all possible future reservations and doubts, to board and seal up all access points through which the machine could be reached and tampered with, and as well, post a guard at each such point. Furthermore, that after the passage of an agreed suitable period of time, We should again inspect said machine and, should all be in order, We should favour him, Orffyreus, with the granting of a document of princely testimony and letters of patronage thereunto relating; the which he has great need of for his own defence and for the rebuttal of unfounded criticism.'

Karl accedes to the inventor's request, declaring that he does so, 'from a love of truth, and from a desire to establish the true facts relating to this most important work. In pursuit of which We shall leave nothing undone . . . This machine emphatically confirmed the claims made for it when, to Our not inconsiderable pleasure, it successfully completed the long awaited month's test, and what is more, did it twice. After the device had been observed in operation for three months by many people, of this district and from elsewhere, of high rank and of low, We finally, on the 12th November of last year, 1717, ordered it to be sealed up and left to run for a fortnight. Then, in person and accompanied by some of Our ministers, we again betook Ourselves, on the 26th November, to the appointed place, and there We opened the seals, which We verified were undisturbed. We carefully observed each and everything We saw and, with Our own hands, We brought to rest the machine We had seen revolving at exactly its original speed. With little effort required, and without the assistance from the inventor We set the machine in motion once more. We then sealed up the machine once more, and all the windows and doors in the room and adjoining areas. Then a full six weeks after all this success, during which period no-one was able to get to the machine, namely on the 4th Day of January of the year 1718, by God's grace newly arrived, We again betook Ourselves to our castle at Weissenstein, whereupon We not only recognised Our impressed seals to be totally inviolate, but also found the Orffyrean Perpetuum Mobile to be continuing just as before in its uninterrupted motion. In addition We found, neither inside the room itself, nor outside the slightest sign of anything suspicious. And so, even though the inventor willingly offered a longer period of running, notwithstanding the fact that the period demanded by his adversary, namely four weeks, had already stretched to eight, We graciously deemed that such an extension was unnecessary for the granting of Our written attestation.'

Karl then enumerates the potential uses for such a machine, mentioning in passing how impressed he had been 'in particular with the large recently built Archimedes screw for raising water'. He has no doubt that once the sale of the machine has been accomplished, a much bigger machine can be built provided that more space is made available and more assistants employed. At the end of the certificate, the Landgrave adds a plea, which is really a royal command.

> 'We . . . request everyone, of whatever rank and position, but especially Our own subjects to refrain from burdening the said inventor, Orffyreus, Our Commercial Councillor, with unjust allegations relating to his unique device, which has not yet become very widely known, and indeed counsel them rather to allow Orffyreus to enjoy the fruits of all such favour, protection and supportive good will as his seemly solicitations and requirements bring him. We are fully desirous of reimbursing anyone who does offer such assistance, as may befit their rank, or of Our own subjects, We will graciously acknowledge their efforts in this cause.
>
> 'Dated Kassel, 27th May, 1718'

It was over four months before Orffyreus received his certificate from Karl, but this was not an oversight but rather because of the Landgrave's other more demanding commitments in the field of international politics. King George I of England, and Elector of Hanover was negotiating the Quadruple Alliance involving contacts between the Czar, Peter the Great of Russia, Charles XII of Sweden, the elected Emperor of the Holy Roman Empire, Charles VI, and various other emissaries from France, Prussia, Saxony, the Polish-Lithuanian commonwealth and the Polish throne. At the centre of much of this diplomatic activity was Karl, the Landgrave of Hesse-Kassel. The reason for his inclusion in much of the political parleying, was partly to do with his geographical position at the centre of the Northern Euıopean nations, thus all traffic tended to pass through his small country - and perhaps more importantly because of his well-trained and well-equipped army of mercenaries, who were much used by the major foreign powers. During the Nordic war, Karl had acted as mediator between Charles XII of Sweden and Frederick I of Prussia. He had had little success, however he had managed to arrange the marriage of his eldest son, Frederick, to Charles XII's sister, Ulrika Eleonora. The acceptance by Charles to Fredrick's marriage to his sister was conditional upon the promise of six thousand Hessian troops to be provided by Karl. These complex negotiations required for their successful conclusion, Karl's presence either in person at various venues, or through his representatives. In either case he needed to be kept well-informed as to his ally's movements and intentions. In addition to contributing to these delicate arrangements, Karl was being approached by representatives of the Jacobites who, in order to improve the chances of securing a line of succession for their master, the exiled James II or, after 1701, James III, were seeking a suitable match for him. Consideration was being given to one of Karl's daughters.

Karl would not have given his blessing to such a union as he was committed to providing the King of England with soldiers in return for subsidies, and George would have taken a dim view of one of his allies permitting a daughter to marry the man who claimed the throne he occupied. On top of all this, on the 27th September of that year, 1717, his sixth son and eventual heir to the Landgraviate of Hesse-Kassel, Wilhelm VIII, married Dorothea Wilhelmina, daughter of Moritz-Wilhem, Duke of Zeitz. The wedding took place at Zeitz and because Karl was there and would have remained for several days afterwards, as a guest of his cousin, Orffyreus' wheel had to await Karl's return before testing began. We can be certain that Karl discussed the Orffyrean wheel with his cousin, after all Moritz-Wilhelm had offered his own patronage to the inventor, and carried out tests to determine the wheel's verity. They probably agreed that there was no need to gather together another group of intellectuals — what was needed was incontestable proof that the wheel could turn continuously for a certain length of time with no sign of diminishing speed. They probably discussed ways of providing this proof and Karl acted upon this advice as soon as he returned.

The request at the end of the document that all Karl's subjects should refrain from bothering Orffyreus, and should instead offer aid when required to do so, is a royal command, and no one would be foolish enough to have ignored it in those days. He even includes people outside his realm, and why not? He may have only been the ruler of a small state, but Karl had powerful friends. His son had married the sister of the young and brilliant King of Sweden; he was providing much needed soldiery to many of Europe's leading sovereigns. No one would risk inviting the Landgrave's wrath by disobeying even so mildly couched a request. It is not surprising therefore, to discover that Gärtner had suddenly become silent about Orffyreus. There was no response either, to Orffyreus' offer of a ten thousand Thaler wager. But there was a full account of the Orffyrean wheel published in the *Acta Eruditorum*. The editor of this respected publication was Johann Burckhard Mencke, who we met earlier; he was one of those present on the 31st October, 1715, at the testing of the wheel at Merseburg.

The article, *An Account of the Perpetuum Mobile of Johann Ernst Elias Orffyreus*, consisted of a brief resume of the story so far, including the actions of the 'Gärtner gang' and their claims that a man was secreted in a hidden chamber, and that he was the real motive force for Orffyrean machine. The article recounts the inventor's various actions, his constant improvements to the wheel, and the fact that he 'silenced his critics, not by words but by deeds'. The previous tests are covered, and mention is made of the illustrious examiners present. It is stated that 'Orffyreus did not attempt to conceal the fact that his machine is set in motion by weights'. The author goes on to 'conclude from circumstantial evidence that the weights were pierced in the middle and attached by connecting springs'.

Later in the above article, the author states that 'certainly, it has been abundantly established that in no way is there any external powering of the machine. Some people find the idea of perpetual motion questionable. They argue that it is possible to build such wheels that can move for some considerable time by means of an internal mechanism, but still need to be rewound, like a clock — and they say that they have constructed such wheels. But these objections have been dealt satisfactorily by Orffyreus'. The report continues with an account of the Karl's examination and test of the machine, including a detailed account of the long duration test, and the many precautions taken to remove all possibility of deception. The account ends with a comment which indicates that Professor Wolff believes that the machine 'is not purely mechanical . . . but that it has not yet been demonstrated that some external subtle fluid may not flow into the moving machine'. He does not know what this 'fluid' might be, but later deliberations suggest that he means gravity.

With the latest test results, the certificates, the favourable publicity and the Landgrave's personal protection, Orffyreus must have thought, yet again, that nothing more was needed. A buyer would appear, the negotiations could begin, and Orffyreus could then get on with his life. He would build a new style of school, 'full of virtue and wisdom', which would be called 'The Fortress of God', and would take in peoples of all nationalities as residents, calling them citizens of God. He would educate them in all the useful crafts of the world, and in the sciences, such as mathematics and mechanics. Devotion to God based solely on the scriptures, would be at the heart of the teaching, combined with a quiet, peaceful and diligent way of life. There was no shortage of potential customers, it was the details, as we shall see, that caused all negotiations to wither and die.

It was now over six years since Orffyreus had first shown his invention to the public. After numerous tests, several moves to new towns, slanderous attacks on his name, the frequent rebuilding of new wheels, the excitement of being given his first real job at Kassel; all this, and yet he was no nearer to a successful sale than he had been in Gera, all those years ago. Orffyreus felt depressed at the way time was passing with no improvement in his fortunes, apart from his success at being called to Kassel, but he had now been a guest of the Landgrave's for over two and a half years. Something further had to be done, and the only possibility was more advertising, he needed something to give impetus to the sale of his machine, and so, in October, 1719, another year later, he published his grandest work. The book, *The Triumphant Orffyrean Perpetual Motion Machine,* was written in both German and Latin, in order to reach as wide an audience as possible. The Latin text would impart a kind of intellectual authenticity to the publication. As a marketing exercise, the book was an outright success. Copies of it travelled all over the world and some can still be found extant in libraries in England, France, Germany, the United States of America and Russia. As well as including copies of all the certificates given in regard to his invention, Orffyreus also added drawings and diagrams, not only of his latest wheel, but also of various applications that it might be utilised for. It is a long document and as full as ever of windy rhetoric, but it also includes his own description of the actual mechanism of the wheel, which although ostensibly giving nothing away, does provide pointers to the way his wheel worked.

Orffyreus presented Karl with a special Christmas gift of the publication. The book contains an account of the history of the perpetual motion, and lists his opponents under four divisions; namely, the scientific world, persons in high authority, the public in general and the press. Which does not seem to leave anyone else, however he does reserve a special place for Karl whose excellent qualities he can not overpraise. He publicly thanks him for 'restoring to me a measure of all the honour and means I had lost in my native country'. He describes how he began to build a new wheel, making it larger than the others and explains that the 'inward structure of the wheel is of a nature according to the laws of perpetual motion, so arranged that certain disposed weights once in rotation, gain force from their own swinging, and must continue their movement as long as their structure does not lose its position and arrangement. Unlike all other automata, such as clocks or springs or other hanging weights which require winding up or whose duration depends on the chain which attaches them, on the contrary these weights are the essential parts and constitute perpetual motion itself; as from them is received the universal movement which they must exercise so long as they remain out of the centre of gravity; and when they come to placed together, so arranged that they can never obtain equilibrium, or the punctum quietus which they unceasingly seek in their wondrous speedy flight, one or another of them must apply its weight vertically to the axis, which in its turn will also move'.

The fact that the wheel was able to turn for eight continuous weeks, night and day, sealed in a room, and under guard, seems to me to be almost incontrovertible evidence of the machine's authenticity. We shall return to the matter of the eight week test, but for now I should like to remind the doubtful reader that the wheel was stopped after this period of time, by the Landgrave, against the wishes of the inventor. Karl felt that the point had been made; if eight weeks was not sufficient length of time to prove that the wheel was not wound up or driven by external sources, how many weeks would be acceptable? The rooms adjacent to the machine-room had been inspected, both above, below and on either side — all possible precautions against deception had been taken. Translocation of the wheel had been performed, weights had been hand-held and examined, close scrutiny of all parts of the wheel had been carried out; and in addition to this we have the Landgrave's statement that he had seen the secret mechanism, and it was genuine.

The book contains an account of Wagner's roasting spit, and even includes Wagner's own diagram of it. This is followed by engravings of the Merseburg wheel, and the Weissenstein Castle wheel with further illustrations of it connected to the Archimedean screw. The book ends with various dedications to Orffyreus, the authors of which, comprise an interesting cross-section of, what appears to be mainly, the local educational establishment.

The first name revealed is that of Lotharius Zumbach de Koesfeld. Karl had appointed him as scientific director of his observatory. Zumbach had gained his doctorate in Leiden, and despite being a Catholic was allowed to continue practising his religion. He was deeply interested in experimental equipment and was responsible for most of the thermometers, hygrometers, mirrors, lenses and telescopes acquired by Karl. Further on we come across the name of Conradus Zumbach de Koesfeld, whom I take to be the former's son or some close relative. Zumbach junior is assigned the title of Mathematical and Medical student. Both contributed short pieces in praise of Orffyreus and his invention. Various names, or in some cases just the initials, follow, but the next one of interest to us is that of J.H. Weber. This was Captain Weber who was dispatched to England in 1716 to obtain information on the Newcomen steam-engine, and is also recorded as being the leader of the surveying team for a projected series of canals which were to enable Karlshafen to become an inland port. He was Karl's chief engineer and he also praised the wheel, and must have had a personal interest in the device in his capacity as an engineer, charged with the task of draining the marshes to create canals.

It may be that in this time of desperation for Orffyreus, he described to Karl how helpful Buchta, the Councillor at Zeitz, had been to him, through his contacts with the now deceased Gottfried Leibniz, and so Karl wrote a letter to his cousin, Moritz-Wilhelm on the 15th January, 1718, asking him to release Buchta from his employment as he wished to take him into his service. Buchta was familiar with the facts and had attended the testing of the Merseburg machine, and he was in constant written communication with the top scientists of the day — having him in close proximity for consultation would have suited Orffyreus. The inventor was advised by Buchta to go into print in two languages in order to reach a wider public. Although German was a major language and French was the language of Court in Europe, Latin was universally taught in every country and was used to lecture in at University, and church services were carried out in Latin. To offer the publication in both German and Latin was a calculated move to attract the attention of all those who enjoyed the benefits of an education, in any country, and who might be in a position to, if not directly buy the Orffyrean machine, at least apply influence in the desired direction.

This apparent stalemate in the saga of the Orffyrean wheel lasted until 1721, when Karl bought and donated to Orffyreus, a house with a garden and a meadow attached to it. It was a stalemate in the sense that Orffyreus wanted to sell his machine and others wanted to buy it — were even prepared to pay the unheard of price of 100,000 thalers for it, but not before they had checked the veracity of the inventor's claims. He, on the other hand, would not allow anyone to see inside the machine until he had all the money in his possession.

In that same year of 1721, two men came to Kassel, who were to contribute two of the most striking pieces of evidence about the Orffyrean machine. They were Professor 'sGravesande and Fischer von Erlach, and their letters to important members of the Royal Society in London add considerably to our knowledge of the legendary machine. But why and how did it come about that less than one month after they were written, those same two letters appeared in a book of famous letters written by the foremost intellectuals of the day, published by a firm of printers in Amsterdam? Later investigations hint at an attempt to blacken the reputation of at least one of the authors. Was this in some way linked to Orffyreus' own problems with Gärtner and his efforts to have the inventor imprisoned? There was also the enduring controversy between Sir Isaac Newton and Gottfried Leibniz about who first invented the calculus. Leibniz may have been dead, but Newton's bitterness had not died with his former enemy, and he was quite capable of adding fuel to the controversy by having private letters published in order to discredit his former adversary. Could these old enmities have been involved in the Orffyrean smear campaign?

Chapter Seven

THE LETTERS

In August 1721, Professor Willem Jacob 'sGravesande visited Karl, the Landgrave of Hesse-Kassel, who had requested his presence during his summer vacation, in order to discuss and evaluate some instruments in his possession. Karl had a room in which were exhibited some of the rarest and most unusual scientific instruments to be seen in Europe, and the reputation that the professor had acquired in the various branches of Physics, made Karl wish to have his opinion on several of the new machines which he would see performing, among them the machine invented by Orffyreus.

'sGravesande was born in the Netherlands, on the 26th September, 1688. He studied law at Leiden University from 1704-1707 and subsequently practised law at the Hague, and in collaboration with others founded the *Journal Littaire de la Haye* in 1713 which contributed significantly to the history of science for some twenty years. He was recognised as a follower and populariser of Leibnizian philosophy.

Early in 1715 'sGravesande was appointed as secretary to the Dutch embassy sent to congratulate George I on his accession to the English throne. This post allowed him to remain in England for some time, and he was introduced to English society and in due course, proposed as a fellow of the Royal Society. He was elected to the Society on the 9th June, 1715 but prior to this date, had attended demonstrations of experiments by J.T. Desaguliers, the Royal Society's Curator of experiments. Sir Isaac Newton was regularly presiding at meetings of the Royal Society and was personally acquainted with 'sGravesande. Desaguliers was also on friendly terms with the young Dutchman.

In June, 1717, 'sGravesande was called to the University of Leiden as Professor of Mathematics and Astronomy. In 1734 he was additionally named Professor of Philosophy. His visits to the Royal Society and his attendance at the weekly lectures and demonstrations held there had a profound impact on him, and he later modelled his own lectures on the examples of Newton in the *Principia* and *Optics*. He also adopted from Desaguliers the idea of demonstrating to his classes the experimental proof of scientific principles, and in the process accumulated an ever growing collection of scientific apparatus. He published his experimental lectures and many subsequent editions followed, corrected and amplified. Desaguliers himself prepared an 'official' English translation from some copies of the original Latin, and these were published in 1720 and 1721. These works were welcomed by both British and German scholars. Before long 'sGravesande had become eminent in an illustrious University. Those seeking a higher education for a profession such as medicine, usually went to Leiden University in the Netherlands. 'sGravesande's lectures proved popular and he attracted hundreds of foreign students each year.

In 1721 he was a rising star, already of formidable reputation, and with his mathematical aptitude and his gift for experimental demonstration, using apparatus designed and largely built by him, worthy of appraising the true value of the Orffyrean machine. When he arrived at Kassel, he found Fischer von Erlach already there.

These two men were the authors of the letters to which I referred earlier. Fischer von Erlach was there to assist in the construction of a steam engine. In 1717, Karl had completed a magnificent park or garden which contained a giant cascade which had, as its source, the famous 'Octagon', an enormous statue of Hercules, looking down from a pyramid, which stood at the top of a landscaped hillside. A powerful pumping engine was needed to pump the water for the cascade to the top of the hill in order for it to issue forth from the feet of Hercules, and flow back down the hill. Karl, perhaps had in mind Orffyreus' wheel for the job, but must have felt that it might not be powerful enough. He may, too, have been remembering all those experiments he had seen Papin carrying out using steam, and, being in constant contact with London, had heard of the successful 'atmospheric' engine of Thomas Newcomen. And of course, he had promised Orffyreus that he would not make use of his invention until it had been sold.

In any case, in 1716, probably during construction of the cascade, Karl sent 'Engineer Captain Johann Heinrich Weber' to England. The entry in Karl's ledger for that year is followed by a recorded expense for 'the construction of a water and fire engine'. Captain Weber, one of those who signed a dedication at the end of Orffyreus' *Triumphans Perpetuum Mobile*, went to England specifically to obtain information about the new steam engine, in order that his employer could have a similar one built to power his cascade. Where would he have gone for his information? Thomas Newcomen or his partner Captain

Savery? Coincidentally, 'sGravesande and Desaguliers had worked together in 1716, on attempting to improve on Captain Savery's original invention and, during a tour of England between 1718 and 1720, Fischer von Erlach had also worked with Desaguliers in a similar capacity. Joseph Emanuel Fischer became, almost certainly, the first to construct a Newcomen engine in continental Europe. Although it is generally believed that the first Newcomen engine was built at Liege and completed in 1721, and another at a drowned out mine at Konigsberg in 1722, there is good evidence that Karl had already begun construction of one as long ago as 1716. However, this may have been a Savery type which was smaller and less powerful.

In his *The Steam-Engine of Thomas Newcomen*, the author L.T.C. Rolt says that drainage in mines was just as big a problem on the continent as it was in Britain, and mines owners there were naturally interested in the Newcomen engine. 'A number of continentals visited Britain and at least one disguised himself as an artisan to discover the secrets of the engine. By 1719 mining interests in both Vienna and Liege were negotiating with English mechanics'. Although the Konigsberg engine was built and the first trials took place in 1722, it took improvements by Fischer in 1724 to make it work properly. It was a new science and the machines although of simple design, took time to perfect. I believe that Karl had his machine built some time before 'sGravesande's and Fischer's visit in 1721, but was unable to get it to work and so called in the experts.

Joseph Emanuel Fischer was the son of Johann Bernhard Fischer von Erlach, a famous Austrian architect, sculptor and architectural historian. The older Fischer was responsible for many of the Baroque style buildings that shaped the artistic style of the Hapsburg empire. In 1696 in recognition for his brilliant work as Court Architect, the father was raised to the rank of Baron.

Joseph's father took care of his son's education himself, and it was extremely comprehensive, including instruction in architecture, drawing, mathematics, physics and the study of antiques. He was also sent to the most important centres of architecture and mathematical sciences in Europe. By the time he was seventeen, his skill at drawing was so far advanced that he was able to give his father's patrons, his future patrons, examples of his work. In 1713 Leibniz had written to the Emperor, telling him how capable the young Fischer was, and as a result Charles VI awarded him a travelling scholarship of 800 florins a year for an indefinite period. Having travelled to Italy and Vienna, he then visited Leiden and Paris. He visited London in 1718 where he worked with Desaguliers evaluating the patents of Savery and Newcomen. Desaguliers was working on improvements to the Savery-Newcomen engine and it was here that Fischer acquired his knowledge of the device. After 1722

Fischer's subsidised educational journey came to an end, and he applied for the position of architect to the court and promised to finish his father's outstanding works, his father being advanced in age and in poor health and unable to finish them himself.

The young Fischer, born in 1693, was only twenty-eight when called to assist Karl. Excellent drawings exist of some of the early steam engines, detailing all the parts, and although none are directly attributable to Fischer, it is inconceivable that he did not make detailed plans of them. No doubt they included accurate measurements of each part plus, of course, any improvements or modifications suggested by Desaguliers while in his employment. These he would have taken with him to Kassel and later to Vienna in 1723, where he caused a great sensation in the city where he built an engine for the Prince of Schwarzenberg, who used it keep the water circulating in his garden and for his fountains.

Joseph Fischer was already at Kassel and had made his own examination of the Orffyrean machine, when Professor 'sGravesande arrived. Fischer sent off his report to Desaguliers, and 'sGravesande, as a respected professor at Leiden University wrote to Newton. Fischer wrote to Desaguliers thus:

> 'I do myself the hour of writing the present letter to mark my esteem for you, and also to give you news of the Perpetual Motion at Kassel, which has so much been recommended to me since I was in London. Although I am very incredulous about things which I do not understand, yet I must assure you that I am quite persuaded that there exists no reason why this machine should not have the name of Perpetual Motion given to it; and I have good reasons to believe that it is one, according to the experiments which I have been allowed to make by permission of His Serene Highness, who is one of the most amiable and gracious princes that I have known in my life; and who had the patience to be present at the trials which I made over a period of two hours. It is a wheel which is twelve feet in diameter, covered with an oil-cloth. At every turn of the wheel can be heard the sound of about eight weights, which fall gently on the side towards which the wheel turns. This wheel turns with astonishing rapidity, making twenty-six turns in a minute, when the axle works unrestricted. Having tied a cord to the axle, to turn an archimedean screw for raising water, the wheel then made twenty turns a minute. This I noted several times by my watch, and I always found the same regularity. I then stopped the wheel with much difficulty, holding on to the circumference with both hands. An

attempt to stop it suddenly would raise a man from the ground.

'Having stopped it in this manner, it remained stationary (and this Sir, is the greatest proof of it being a Perpetual Motion), I commenced the movement very gently to see if it would of itself regain its former rapidity, which I doubted, believing as they said in London, that it only preserved for a long time the impetus of the impulse first communicated. But to my astonishment I observed that the rapidity of the wheel augmented little by little until it had made two turns, and then it regained its former speed, until I observed by my watch that it made the same twenty-six turns a minute as before, when acting freely; and twenty turns when it was attached to the screw to raise water.

'This experiment, Sir, showing the rapidity of the wheel augmenting from the very slow movement I gave it, to an extraordinary rapid one, convinces me more than if I had seen the wheel moving for a whole year, which would not have persuaded me that it was perpetual motion, because it might have diminished little by little until it ceased altogether; but to gain speed instead of losing it, and to increase that speed to a certain degree in spite of the resistance of air and the friction of the axles, I do not see how any one can doubt the truth of this action. I then turned it in the opposite direction, and the wheel produced the same effect. I examined the bearings of this wheel to see if there was any hidden artifice; but was unable to see anything more than the two small bearings on which the wheel is suspended at its centre.

'His Highness, who possesses all the qualities that a great prince should have, has always had consideration for the inventor, and will not use the machine in any way for fear of the secret being discovered before the inventor had received a reward from foreigners. His Highness, who has a perfect understanding of mathematics, assured me that the machine is so simple that a carpenter's boy could understand and make it after having seen the inside of this wheel, and that he would not risk his name in giving these attestations, if he did not have knowledge of the machine.

'I said to his Highness that I had no doubt a company might be formed in London to purchase the secret. The prince would be exceedingly happy if such a company would consign into his or other hands £20,000 for the inventor, then the machine should be examined and the secret communicated. If the movement were found to be a perpetual one, the £20,000 would be given to the inventor; and if not, the money would be returned. This would be stipulated by proper legal documents. I told His Serene Highness that no one could institute such a company better than yourself, for you are always working for the instruction of the public. Consider under what obligation you would place the most enlightened nation of Europe, if you procured for it the knowledge of the principle of this perpetual motion, as by that means you would discover an infinity of beautiful inventions that are now unknown. As I shall not remain here long, I must beg you to correspond with Mr. Roman, Superintendent of His Highnesses buildings. He will show all your letters to the Prince, and will come to an understanding with you touching this matter, which well merits your highest consideration, as it is unwise to leave this treasure buried. Will you also please communicate with your friend Sir Newton, and tell him my opinion of the machine. I hope that you will soon hear from our friend Mr. 'sGravesande of Leiden, who I hope to see soon, as he is on a small tour and will present his compliments to His Royal Highness; His Highness has sent a letter via Mr Roman, saying that he wishes to see him here.'

Fischer's letter leaves no doubt that he was convinced that Orffyreus' machine was genuine; unfortunately he was unaware or could not conceive that his friend, Desaguliers, might find that his request for assistance in spreading the word about the Orffyrean machine might conflict with his own plans for developing the Savery-Newcomen engine. This first letter was followed immediately by 'sGravesande's similar one to Sir Isaac Newton:

'Doctor Desaguliers has doubtless shown you the letter that Baron Fischer wrote to him recently, about Orffyreus' wheel, which the inventor asserts is a perpetual motion. The Landgrave, who is a lover of the sciences and arts, and who neglects no opportunity to encourage the several discoveries and improvements that are presented to him, was desirous of having this machine made known to the world, for the sake of public utility. To this end he engaged me to examine it, wishing that, if it should be found to answer the pretensions of the inventor, it might be made known to persons of greater abilities, who might deduce from it those services which are naturally to be expected from so singular an invention. You will not be displeased, I presume, with a circumstantial account of this examination. I send you therefore the details of the most particular circumstances observable on an exterior view of the machine, concerning which the sentiments of most people are greatly divided, whilst almost all the mathematicians are against it. The majority maintain the impossibility of a perpetual motion, and hence it is, that so little attention has been paid to Orffyreus and his invention.

'For my part, however, though I confess my abilities inferior to those of many who have given demonstration of this impossibility — yet I will communicate to you the real sentiments with which I entered on the examination of this machine. It is now more than seven years since I discovered the erroneous reasoning of those demonstrations, in that, though true in themselves, they were not applicable to all possible machines; and have ever since remained perfectly persuaded that it might be possible to demonstrate that a perpetual motion involved no contradiction. It seemed to me that Leibniz was wrong in laying down the impossibility of perpetual motion as an axiom. Notwithstanding this persuasion, however, I was far from believing Orffyreus capable of making such a discovery, looking upon it as an invention not to be made (if ever) till after many other previous discoveries. But since I have examined the machine, it is impossible for me to sufficiently express my astonishment.

'The inventor has a turn for mechanics, but is far from being a profound mathematician, and yet his machine has something in it prodigiously astounding, even though it should be an imposition. The following is a description of the external parts of the machine, the inside of which the inventor will not permit to be seen, lest anyone should rob him of his secret. It is a hollow wheel or kind of drum, about fourteen inches thick, and twelve feet in diameter; being very light as it consists of several cross pieces of wood framed together; the whole of which is covered over with canvas, to prevent the inside from being seen. Through the centre of this wheel or drum runs an axle of about six inches diameter, terminated at both ends by iron bearings of about three-quarters of an inch diameter upon which the whole thing turns. I have examined these bearings, and am firmly persuaded that nothing from without the wheel in the least contributes to its motion. When I turned it but gently, it always stood still as soon as I took my hand away. But when I gave it any tolerable degree of velocity, I was always obliged to stop it again by force; for when I let it go, it acquired in two or three turns its greatest velocity, after which it revolved at twenty-five or twenty-six times a minute. This motion it preserved some time ago for two months, in an apartment of the castle; the door and windows of which were locked and sealed, so that there was no possibility of fraud. At the expiration of that time, His Serene Highness ordered the apartment to be opened, and the machine stopped, lest, as it was only a model, the parts might suffer by so much testing. The Landgrave being, himself, present during my examination of this machine, I took the liberty to ask him, as he had seen the inside of it, whether, after being in motion for a certain time, some alteration was made in the component parts; or whether one of these parts might be suspected of concealing some fraud; on which His Serene Highness assured me to the contrary, and that the machine was very simple.

'You see. Sir, I have not had any absolute demonstration, that the principle of motion which is certainly within the wheel, is really a principle of perpetual motion; but at the same time it cannot be denied me that I have received very good reasons to think so, which is a strong presumption in favour of the inventor. The Landgrave made Orffyreus a very handsome present to be let into the secret of the machine, under an engagement nevertheless not to discover, or to make any use of it, before the inventor has procured a sufficient reward for making his discovery public. I am very well aware, Sir, that in England only, the arts and sciences are so generally cultivated as to afford any prospect of the inventor's acquiring a reward adequate to this discovery. He requires nothing more than the assurance of having it paid him when his machine is found to be really a perpetual motion; and as he desires nothing more than this assurance till the construction of the machine be displayed and fairly examined, it cannot expected he should submit to such an examination before such an assurance be given him. Now, Sir, as it would be conducive to public utility, as well as to the advancement of science, to discover the reality or the fraud of this invention, I conceive the relation of the above circumstances could not fail to be acceptable . . . '

There is no record of any reply to these two letters. It is possible that they were replied to, but none has surfaced and it is fair to assume that none exists. There was a response of a kind. In the above letter 'sGravesande remarks that he has not seen the impossibility of perpetual motion demonstrated; within one month Desaguliers gave a public lecture intended to demonstrate just that - and he had the lecture published later. More ominously and mysteriously, these letters were published in a book of letters on scientific and intellectual matters, again within only one month of their being *written*, by a firm of publishers in Amsterdam. Both Sir Isaac Newton and Dr. Desaguliers seem to have, at best ignored the letters as unworthy of consideration, but did they also deliberately attempt to embarrass 'sGravesande by having his letter published? The professor himself, was mystified as to how the letters came to be published. He said later, that they were never intended for public consumption. One is forced to conclude that the letters were forwarded to the publisher by the recipients for some secret motive. But why? It cannot have been done to advertise Orffyreus' wheel, because that would imply that the recipients of the letters supported the possibility of perpetual motion. Desaguliers did *not* support the theory of perpetual motion, in fact, as I mentioned, during that very month he gave a lecture at the Royal Society in which he attempted to demonstrate the impossibility of perpetual motion. A full record of this lecture exists, because Desaguliers published it in a book of other lectures he had given over a period of time, entitled *A Course of Experimental Philosophy* in two volumes in 1745.

He prefaced his lecture with the following remarks:

> '14 [70 – Pretenders to perpetual motion, and those who promise greater effects by machinery than is conformable to the reciprocal proportion between the intensities of the powers and weights, and their velocities.] ' (This clumsy title is a statement to the effect that this lecture is aimed at people who claim that perpetual motion is possible by means of weights. At the same time Desaguliers is throwing doubt on their claim by implying that such a notion is impossible because the falling weights and their velocity would be insufficient to produce the effects claimed by the adherents of the perpetual motion theory). He continues thus:
>
> 'About the year 1720 and 1721, the late John Rowley, mathematical instrument maker, talk'd so much of the wheel which he had seen at Hesse-Kassel (which he believed to be a perpetual motion, along with a great many persons in that country) that besides the common herd of Perpetual Motion men, which every age affords, some very ingenious men made an attempt that way, and were countenanc'd in it by some great mathematicians, who, when the scheme was laid before them, declar'd they knew no reason why it should not do. But as I always declar'd against all projects tending that way, I was desir'd at that time to publish my reasons why the thing seem'd impossible or impracticable; which I did in the *Philosophical Transactions* (No369) in such a manner as might dissuade people at first from any such attempts, in which so much time and money have been lost. I have here printed the whole account again.'

He then reproduces the lecture which he originally gave in September, 1721. There is no need to reproduce the lecture, which is sound enough up to a point, with several diagrams and fairly unarguable statements of fact. What is interesting is not so much the timing of the lecture which might perhaps have benefited from the publication of 'sGravesande's letter, but rather, the single mention of the name John Rowley, 'who talk'd so much of the wheel he had seen at Hesse-Kassel'. I have only ever been able to find one other reference to John Rowley in connection with the Orffyrean wheel, and that was an equally brief mention by Samuel Johnson in *The Adventurer*, (no 99) dated Tuesday, 16th October, 1753. In it Johnson says,

> 'They who find themselves inclined to censure new undertakings, only because they are new, should consider, that the folly of projection is very seldom the folly of a fool; it is commonly the ebullience of a capacious mind, crowded with variety of knowledge, and heated with intenseness of thought; it proceeds often from the consciousness of uncommon powers, from confidence of those, who having already done so much, are easily persuaded that they can do more: when John Rowley had completed the Orrery, he attempted the perpetual motion; when Bole had exhausted the secrets of vulgar chemistry, he turned his thoughts to transmutation.'

John Rowley had seen the wheel at Hesse-Kassel. When he returned to England, he was so convinced that what he had seen was genuine that he spent the rest of his life attempting to duplicate the machine. Who was John Rowley? His only apparent claim to fame was the connection with the Orrery, a device invented to demonstrate the motion of the planets and the sun around the earth. We shall return to Rowley; he is inextricably linked to the Orffyrean wheel through various channels — Newton, Desaguliers, King George I, the Royal Observatory at Greenwich, Sir John Vanbrugh the architect and playwright - and the most exclusive club in London, St. Luke's. At the moment Desaguliers merits our attention; why was he so public in his utterances against the idea of perpetual motion? Why was Newton silent about perpetual motion, after all as we saw earlier, he had toyed with the idea of perpetual motion, even making sketches of how he thought it might be achieved?

Desaguliers had worked as an assistant to Newton at the Royal Society for a number of years. It is hard to say if they were friends, as neither were particularly likeable people. Desaguliers was born on the 13th March, 1683, the son of a Protestant Pastor. Like many Huguenots, his father had been forced to flee to England when the Edict of Nantes was revoked in 1685. After his father's death John Desaguliers obtained a BA degree at Oxford University in 1709, followed by an MA in 1712. He took up residence in Channel Row, Westminster and gave lectures on Hydrostatics, Optics and Mechanics. He was elected to the Royal Society in 1714, and was invited to become their demonstrator and curator. Sir Isaac Newton thought highly of him and employed him as curator of experiments. He lectured in front of King George I, in 1717, and a year later received his degree at Oxford as Bachelor and Doctor of Laws. He is said to have been the first to have delivered learned lectures to the public in general. His lectures at his home were widely attended and were made attractive by the inclusion of experiments. Yet despite this he was always short of money. His lectures, though popular brought insufficient money, and his last years were spent in want, and obscurity with bouts of jealousy and insanity. He died in 1749, some say in poverty. After all this time it is difficult to judge his character accurately, but there are clues. If he was jealous in his old age then that trait was probably present in his prime. He was guilty of self-aggrandizement, and he had strong reasons for rejecting any consideration of the Orffyrean machine, as we shall see.

Newcomen's early researches and experiments which led to a successful machine and his eventual partnership with Captain Savery are well described in L.T.C. Rolt's book, *The Steam Engine of Thomas Newcomen.* Suffice to say that in 1712, the first positively recorded working Newcomen engine was that mentioned by Martin Triewald, a young Swedish engineer, who came to England in 1716, aged twenty-five to study English industrial methods, and stayed ten years. He wrote,

> '. . . Later on Mr. Newcomen built the first fire-engine in England in the year 1712 . . . at Dudley Castle, in Staffordshire . . . The fame of this excellent pumping engine soon spread across England and many people came to see it, both from England and from foreign nations. All of them wanted to make use of the invention at their own mines and exerted themselves to acquire the knowledge needed to make and erect such a wonderful engine, but the inventors, Newcomen and Calley (Newcomen's original assistant) were exceedingly jealous and very anxious to preserve exclusively for themselves and their children, the knowledge of making and operating their invention, which had cost them such unprecedented toil. Even the Spanish ambassador to the Court of St. James's who came all the way from London with a large suite of foreigners, to see the engine, was not even allowed to enter the engine house however large a reward he offered and had to return very dissatisfied without having seen anything but the wonderful results, that this small engine was able to produce.'

This first engine, of which a modern working copy has been constructed close to the site of the original, was able to work at a rate of twelve strokes a minute, pumping ten gallons of water per stroke from a depth of over 150 feet — no wonder there was so much excitement. It is interesting to observe exactly the same reticence on the part of the inventors in displaying the working parts of their engine as that shown by Orffyreus.

Thomas Savery had also invented a steam engine but one that worked on an entirely different principle to Newcomen's. He was a prolific inventor, and was granted, on the 25th July, 1698, a patent for 'raising water by the impellant force of fire'. This patent was originally granted to run for fourteen years, but was later extended by an additional twenty-one years. The principle behind it differed from Newcomen's because there were hardly any moving parts, it was smaller, and appeared less wasteful. In June 1699, Savery had demonstratcd to the Royal Society an improved version of his engine. In 1705 Savery was elected a Fellow of the Royal Society and became active in its affairs. Papin took a deep interest in Savery's engine and in 1707 presented a paper to the Royal Society in which he offered some improvements to it. For all this, the Savery engine had its limitations. It proved impossible to lift by suction more than twenty feet, which was hopelessly inadequate for mining purposes. The forced lift which could be brought into play on the upper side of the machine also proved to be too weak due to the low steam pressure which the boiler could withstand.

As L.T.C. Rolt puts it, 'Desaguliers belonged to that scientific school which refused to credit that a practical man without scientific training could achieve anything except by chance and good luck. Therefore, according to him, the steam engine was simply one long, phenomenal run of luck'. Desaguliers confirms this view of himself by writing, 'If the reader is not acquainted with the History of the several improvements of the fire-engine since Mr Newcomen and Mr Calley first made it go with a piston, he will imagine that it must be owing to great sagacity, and a thorough knowledge of philosophy, that such proper remedies for the inconvenience and difficult cases mentioned were thought of; but here has been no such thing; almost every improvement has been owing to chance'. What an amazingly arrogant view of Newcomen and Calley's hard won success! I had already come to the same conclusion about Desaguliers before I read L.T.C. Rolt comments above. It shows in his sweeping statements about the Orffyrean wheel. He ignores all the reports by competent witnesses; he disregards the evidence; he looks no further than the inventor's claim that the machine is a perpetual motion machine, merely remarking that such a claim is impossible therefore implying that the man must be a fraud.

Savery was ex-army, of good education and therefore admissible to the Royal Society. Because Savery held the patent for raising water by fire,

Newcomen had no choice but to join forces with him in order to be allowed to operate his own engines. Desaguliers continued to work on the Savery engine trying to improve it, ignoring the fact that it had not got one tenth the power of a Newcomen engine. Was he somehow trying to improve the Savery engine to such an extent that it would supersede the Newcomen engine, thus wiping out the embarrassment of having to explain away the ill-educated Newcomen success? L.T.C. Rolt says, 'Curiously, despite the evident limitations of Savery's system, it continued to attract scientists and engineers long after the death of the inventor in May, 1715, and long after Newcomen's engine had proved itself beyond dispute'.

Professor Allemand, who wrote the biography of professor 'sGravesande states that Desaguliers recommended Fischer to Karl, as 'a very good Mechanician', and this is why he was called to Kassel to assist in setting up the steam engine required to pump the water for the cascade. Captain Weber's engine may have been the Savery type, as they were easier and cheaper to construct, but would have proved useless for the purpose that Karl had in mind. Doubt has been expressed as to whether a Newcomen type engine was ever built at Kassel, but in a reported conversation with Fischer in 1727, he actually claimed that he did build an engine there, and it is certain that Fischer did build the engine in Vienna in 1723 for the Prince Schwarzenberg, so he had the necessary skills.

Fischer, then, wrote to Desaguliers, a man who was jealously promoting his own pet project, the Savery pump. He had ignored Newcomen's superior engine, and in the same way he ignored the Orffyrean wheel. Newton, the other recipient of the latest news on Orffyreus' wheel, held a deep seated hatred for Leibniz that continued despite the death of his enemy some years earlier. Knowing as he undoubtedly did, that Leibniz had become fascinated with Orffyreus' machine, Newton may well have seen a way of injuring Leibniz' reputation. The cause of the dispute lay in the past over who had the right to claim the discovery of the new infinitesimal calculus; today's differential and integral calculus. Leibniz had published an outline of his differential calculus in 1684, at which time none of Newton's mathematical writings had been officially published. Newton was provoked into a war of words that continued after the death of Leibniz in 1716. Newton asserted that if Leibniz had been first in print, it had been with the knowledge of what Newton had done previously. He claimed that in the first place he had made his initial discoveries in the calculus nearly twenty years before Leibniz's publication; and secondly, he had shared these innovations with Leibniz and others during those intervening years. Newton was a difficult man and harboured interminable grudges against people. This hatred coloured all aspects of their relationship. Even though Leibniz had been dead for six years, Newton continued to

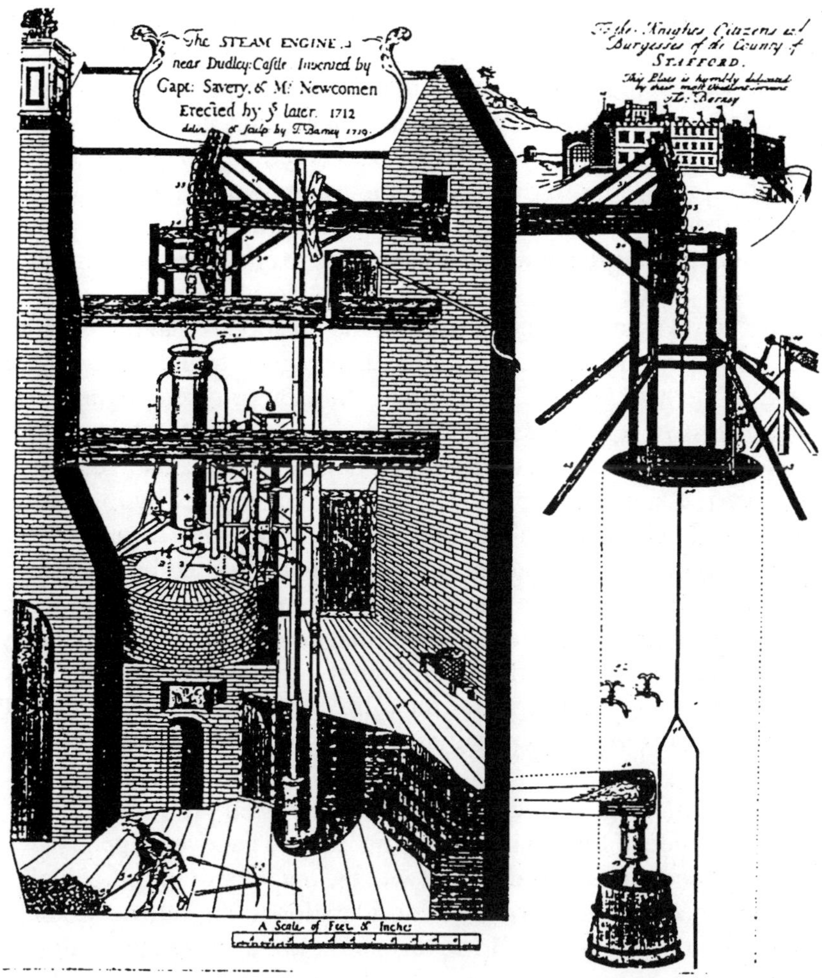

Dudley Castle Newcomen Engine

This is the earliest depiction of a Newcomen engine known. It is of the famous Dudley Castle engine, and the castle can be seen in the top right hand corner of the illustration. There is currently a full size copy of the engine working at the very spot it is believed the original occupied.

allow his supporters to savage Leibniz's reputation. If Leibniz was a supporter of Orffyreus, then any action which might bring discredit to Leibniz's reputation was acceptable, despite the fact that the new perpetual motion machine could be lost in the process. There was much doubt about the validity of Orffyreus' claims, but such expert witnesses could not be ignored, however printing the claims of those witnesses without comment might be sufficient to add fuel to the implication that Leibniz's genius was flawed.

One would imagine that it would be necessary to obtain permission from the author before a letter could be published. Not so in this case — and poor Professor 'sGravesande found himself in the position of having to publicly defend his words. There was also the disparaging figure of Desaguliers hovering in the background, encouraging Newton or discouraging him according to his own perceived requirements. In August of that year (1721) he gave a lecture to demonstrate the impossibility of perpetual motion. The lecture was, he claimed, a response to all the talk engendered by John Rowley, who was sure that he had seen a genuine perpetual motion. The publication of the 'sGravesande letter would do no harm for his proposed lecture.

Professor 'sGravesande's examination of the machine at Kassel was thorough and, in fact, upset Orffyreus so much that the following day he destroyed his machine and wrote upon the door of the room in which the machine was kept, a verse blaming the destruction on 'sGravesande. Orffyreus refers to this verse the following year in a panegyric written for Karl. He says,

> 'On the 16th August, 1721, at Castle Weissenstein, Herr Professor 'sGravesande of Leiden in Holland had somewhat overstepped the limits of decorum in his dealings with the inventor of the perpetual motion machine. He gave little consideration to the certificates of attestation for the wheel. However once he had accepted the reality of the work, he was overcome with admiration. On the door of the room in Weissenstein Castle, where the perpetual motion was kept, the following verse is written.,
>
> 'Here's where the Perpetual Motion ran,
> Invented by Orffyreus.
> Many spectators saw the sight,
> Among them, Herr Hate and Herr Jealousy!
> What was proved true will ever-last
> But Weissenstein saw what came to pass,
> The destruction of the wheel!
> Many know why I smashed my wheel.

It was the German nation's fault
But Karl has done his best to heal.
Let thanks be given to God and to Karl,
Praise the Landgrave, within whose care
Turns the wonder-wheel to Karlshafen.'

This short verse, difficult to translate, mentions Karlshafen because that is where Orffyreus moved to, courtesy of Karl 's generosity. He bought the inventor a house and garden in the nearby town of Karlshafen. This fact should be born in mind when considering the reasons for destroying the wheel. The reader will remember that Orffyreus had a propensity for smashing his wheel prior to moving house. His suggestion that it was the discourtesy of Professor 'sGravesande that caused him to destroy his machine again does not really stand up, especially when we re-read his words above, i.e. '. . . once he had accepted the reality of the work, he was overcome with admiration.' If the inventor knew that the professor admired his machine, even if at first he queried its authenticity, surely he would have been pleased with the result; and would not have wished to destroy it.

The certificates may have proved the machine's worth, but Professor 'sGravesande had a reputation to maintain; he was not going to recommend a machine until he had made a thorough examination of it, and he was not going to accept the word of another who, perhaps, he did not know. Once he had satisfied himself as far as possible as to the machine's validity, Professor 'sGravesande set about establishing interest in the machine within the circle of scientists in England. He started at the top, as it were, by addressing his letter describing the wheel, to Sir Isaac Newton; and as we saw, unfortunately, a letter destined for the private perusal of a top scientist ended up published in a book, scarcely a month later.

This caused the professor considerable embarrassment and he was compelled to issue an official explanation in defence of himself. The defence is long and there is no need to reproduce the whole text. It begins thus,

> 'About eight months ago, by order of His Serene Highness the Landgrave of Hesse, I was called upon to examine the effects of a machine at Kassel, which the inventor (Orffyreus) claimed to be a perpetual motion. He carefully hid the interior, requiring a certain sum of money once the machine had first been examined and recognised by mathematicians to be, what in mechanics is called, perpetual motion. I was so forcibly struck by what I saw and what I heard, as to be unable to call it in question. I was obliged to look

on the machine as one of the most beautiful inventions that I have ever known, and only to consider the truths of the effects before me. I wrote to Mr Newton all I had observed about it, which having been published, fault has been found with the following statements:-

"I do not believe perpetual motion to be a contradiction." -
"the demonstrations given on its impossibility do not appear to me to be applicable to all machines," — and lastly
"I find it probable that the machine at Kassel may be a perpetual motion."

The whole problem is based on the first of these remarks; if that were proved, the remaining ones would not embarrass me much. It was probably too bold of me to venture this without proof. In this I quite agree, and I would never have committed the fault had my letter been written for public consumption. I have always been so careful not to advance a proposition without proof, that I have never yet declared what I thought on perpetual motion, foreseeing the judgement that all mathematicians would pass on anyone who turned aside from the generally received opinions. Having considered what I owed to truth, after having seen the machine at Kassel, I was induced to give to Mr Newton my opinion on it, and at the same time to say what I thought of the proofs of the impossibility of perpetual motion. Now I have to justify myself before the public.'

Professor 'sGravesande continues by defining exactly what he means by perpetual motion. He describes the current explanations and disagreements between scientists as to whether perpetual motion is possible or not, quoting from Leibniz and Papin. He disagrees with Leibniz's view that it is impossible. He does, however agree with Leibniz in one thing, he says,

'The question of the possibility or impossibility of perpetual motion then, seems to me to be of little consequence; but it is desirable that the strong belief that mathematicians have regarding its impossibility should not prevent their paying serious attention to such a machine as the astonishing wheel of Kassel; a wheel having the principle of internal movement, and which is moved by the slightest effort, turning either way without any necessity to reverse the moving power; lastly, after making some millions of surprisingly rapid turns, continues in its motion until stopped by a strong effort

> of the arm. It appears to me that such a machine merits some praise, even if it does not satisfy all its inventor's representations. If it is perpetual motion, he deserves the recompense he asks; if not, the public may, nevertheless, here discover a beautiful invention.'

'sGravesande's conviction that the Kassel Machine is genuine, even if it is not actually a perpetual motion machine, is impressive. At this point I must draw the reader's attention to something rather curious. Almost all accounts of the Orffyrean wheel assume some kind of fraud — or leave the ending open as a tantalising mystery. But they make one extraordinary error concerning Professor 'sGravesande. During my research on this subject I have come across no less than three accounts written over a period of some fifty years or more, which all use the same argument in support of their author's contention that Orffyreus was a fraud. They explain away Professor 'sGravesande's belief in the wheel in the following manner. I will quote from the latest account which uses very similar words to the two previous ones. Arthur W.J.G. Ord-Hume's book is called *Perpetual Motion – The History of an Obsession*. After describing the examination of the wheel at Kassel by 'sGravesande, the author remarks,

> 'The Professor certainly seems to have had some measure of faith in the wheel and the demonstration of its ability to turn without apparent external force. We ought not to forget, though, that it may have proved easy to dupe an honest old man whose confidence in humanity was probably unbounded . . . It is not recorded whether the aging academician ever received a reply to his letter to Sir Isaac Newton.'

As this argument attempted to cast doubt on the professor's competence to decide whether the machine was valid or not, it was important to check his age. If he *was* elderly, then his capacity as a valuable witness was seriously impaired. In fact, this 'honest old man', this 'aging academician' was born on the 26th September, 1688, which means that at the time of the Kassel examination he had reached the grand old age of thirty-three! He lived for over another twenty years, dying in 1742. In 1730 he was described as one of the great luminaries of scientific experimentation at Leiden University, possibly the foremost University of the age. His lectures drew the biggest crowds of students. This does not sound like a man of failing intellectual powers, and he was certainly one who could form an opinion based on the evidence in front of him, which others could rely on as accurate. He would not have upset

Orffyreus so much if he had not had the courage to ask the most searching questions concerning the Kassel wheel. The argument that he was old and gullible is invalidated.

In a scientific publication, *Neue Zeitungen von Gelehrten Sachen*, which was published by the Mencke family in Leipzig in the April issue of 1722, page 344 it said,

> 'In Kassel, Mr Orffyreus, who is Commercial Councillor to His Highness the Landgrave, has destroyed his Perpetuum Mobile that, prior to this, stood at the Landgrave's Castle of Weissenstein. He did this after he had suffered a serious illness, and after he had received a house in Karlshafen. The house was bought and donated to him by His Serene Highness the Landgrave of Hesse-Kassel. Karlshafen is a newly constructed commercial town on the rivers Diemel and Weser, and it was previously called Sieburg.'
>
> 'Mr. Orffyreus plans to build a new, much larger machine there, and to extend its use for some beneficial application. In addition he has published a booklet about it under the following title: *The Triumphant Orffyrean Perpetuum Mobile. . . Kassel, 1722*'

Orffyreus, excellent self-publicist that he was, smashed his wheel again and got himself into the newspapers. The fact that his public generally regarded his actions as bordering on madness either escaped him, or he did not care. What is curious is what exactly prompted the move in the first place. Boredom on the inventor's part may have been a factor. He had, in the past, had a wanderlust which refused to allow him to settle in one place for long. As well he preferred the countryside to the towns. Another reason may have had something to do with Karl's age — he was sixty-eight. He had carried the expense of supplying materials for the building of the Orffyrean wheel; the provision of apartments in his Castle; the establishment of a salaried position for Orffyreus; the eight week test and all that that entailed and a constant stream of visitors to see the machine. All this for a period of six years, perhaps longer than he had originally anticipated. It is known that Karl suffered from a gradually deteriorating mental condition, possibly Altzeimer's disease, but as this did not become apparent for another five years it is unlikely to have been a factor in the decision to move Orffyreus to Karlshafen.

Karl's wife, Marie Amalie, who had borne him fifteen children during their thirty-eight years of marriage, had died in 1711, and towards the end of his life he had had relationships with two women at his court. The first was the Marquise de Langallerie, who had accompanied her husband to Kassel in

1713. She had stayed behind when her husband had moved on after a year. He was Philip de Gentils. the French adventurer described earlier. Shortly after her husband had left, she entered into a close relationship with Karl, so close in fact, that it was generally assumed that the two sons she bore in 1714 and 1716 were his. In 1716 the Marquise's annual pension was six hundred thalers (£120), but by 1728 Karl had raised it to two thousand thalers (£400). At today's rate somewhere in the region of £30,000 p.a.!

In 1722 she acquired a rival for Karl's affections, namely the attractive Christine von Bernhold. Her introduction to the court resulted from the good military service her father had earlier performed. A special court position of a ceremonial nature was created for her, bearing the extraordinarily high annual remuneration of four thousand thalers (£800). Karl's sons became increasingly concerned about the excessive influence the two women seemed to be having on their father. The Landgrave's mental and physical powers had since about 1726 gradually begun to decline and when it was learned that he had bestowed on the two women jewellery, land and property totalling 100,000 thalers (£20,000) his sons had cause for concern. Prince Wilhelm in particular, who from about 1727 onwards spent more and more time at court on account of his father's increasing indisposition, came to detest the Marquise de Langallerie especially because of the power she wielded.

The rivalry between the two women increased, but Karl managed to keep them both 'sweet' by increasing concessions to them, thus both women, with their separate circle of hangers on, continued to have influence. The Landgrave's three surviving sons remained loyal to their father, but felt that they had to draw the line when the War Chest, no less, was raided to secure the continued favours of the two women. Nevertheless, Langallerie in particular remained powerful, almost to the time of Karl's death on the 23rd March, 1730.

I believe that the Marquise de Langallerie was behind the move to get Orffyreus out of the Castle. Exactly what her motives were we can only surmise, but a clue to the reasons for her actions will be examined shortly. Certainly she was disliked by Karl's sons, and no one else had the power and influence over the Landgrave that she had. Of course it may have involved Karl's other mistress, the Countess Bernhold, but she has never been regarded as sinister in the way that the Marquise has. One reason for her wanting Orffyreus out may have been due to his self-proclaimed piety. He may have expressed disapproval of her position and influence over Karl. Also she was a Catholic which was anathema to Orffyreus. The 100,000 thalers given by Karl to his two mistresses, matches exactly what Orffyreus was asking in payment for his machine, perhaps he complained that that money could have been better used in purchasing his machine. He was not above loudly denunciating those of whom he disapproved and the Marquise probably made a bad enemy.

Immediately following the news of Orffyreus' move to Karlshafen, Gärtner reared his head again and caused or had caused, the inventor's arrest on a trumped up charge. He may even have been a shadowy figure behind the scenes exerting his baleful influence over the Marquise.

Long before this though, at about the time that Orffyreus first came to Kassel, the Czar of Russia, Peter the Great, was in Carlsbad taking a cure in the waters there, and to witness the marriage of his son, Alex to Princess Charlotte of Wolfenbüttel. While on his travels through Europe, Peter paid a visit to Dresden where, among others he met Andreas Gärtner, who on this occasion was extolling the virtues of a new machine he had just contrived, designed to carry people or objects from floor to floor; in other words an elevator. It was during this trip that Peter first heard about the Orffyrean perpetual motion.

Chapter Eight

PETER THE GREAT — CZAR OF RUSSIA

Peter the Great left Russia in 1697 to tour Europe, the first Czar ever to leave his own country. His aim in travelling through Europe was to learn better how to govern his own domain. He visited London and stayed for six months studying everything from ship-building to watch-making. The Czar had a deep and lasting fascination with things of a technological nature, and on one visit to a watchmaker he stopped to buy a pocket watch, and stayed to learn how to dismantle, repair and reassemble the intricate mechanism.

Peter's chief aim was to civilize Russia. He established schools, universities, libraries, and new Naval Academy. The efforts in education had the effect of stimulating a demand for books, and scientific instruments of all sorts. The importation of thousands of books was ordered, even the buying of complete libraries. An Academy of Science was planned, and Gottfried Leibniz and Christian Wolff were contacted for their advice. Wolff was eventually requested to come and live in Russia and assist in setting up the Academy of Sciences. In 1721 and 1722, Peter's librarian, J.D. Schumacher, toured much of Europe attempting to make contact with eminent foreign scholars, buying books and scientific instruments for the Czar's new Academy. On his return to Russia in 1722, Schumacher brought with him almost 600 books. He reappears as one of Peter's representatives negotiating with Orffyreus to purchase his perpetual motion machine.

All the ideas for the development of the Academy of Sciences came from Leibniz's fertile mind, and he was aided by Wolff. Although he never went to Russia, Leibniz sent copious quantities of memoranda, particularly regarding the Academy of Sciences, reform of justice and government, and various scientific and geographical projects.

It was in these early correspondences that Peter became aware of Leibniz's interest in the Orffyrean perpetual motion machine, and decided to buy the secret of the machine for the benefit of Russia. Being unable to examine the machine himself, a fact that must have irked him, he sent his representatives to negotiate with the inventor. It was not until 1725 that an opportunity to personally examine the Orffyrean machine arose, but unfortunately for him and Orffyreus and perhaps, the rest of us, he was overtaken by events and died before being able to complete the journey.

Correspondence exists allowing us to follow the train of events. On the 22nd July, 1716, Wolff wrote to Leibniz,

> 'Two men have arrived here in Halle and are intending to travel to Merseburg in order to buy Orffyreus' machine on behalf of the Czar of Russia. They are aware that Orffyreus is about to move to be under the protection of the Landgrave of Hesse-Kassel. One of them spoke with me and asked me to accompany them when they visit Orffyreus in order to assist in arranging an agreement with him.'

On the 9th August, 1716, Wolff wrote again to Leibniz,

> '. . . the man who visited me and was sent to Merseburg by the Russian Monarch was a German, but I do not know his name or his rank. He told me of another man who was already in Merseburg, and was also a German, and Chancellor of the Exchequer. He was used by the Czar to expedite some affairs with Germany. His name is also unknown to me. Last year, the Commissioners and their guests had to be satisfied with just the translocation of the machine to another support. The inventor was very anxious to repudiate Gärtner's accusations that his machine was moved by a cord wrapped around the axle, but he was unwilling to allow any other test at the time, because there was no time to do it due to the holiday and the Commissioners were late and only began their enquiry at noon.'

Wolff is referring to the earlier test at Mersburg, prior to Orffyreus' move to Hesse-Kassel. The two men mentioned in the letter were Andrei Ivanovich Ostermann and Baron Peter Shafirov.

Shafirov was from a Jewish family in Smolensk, Poland. He had been made a Baron in 1710, Russia's first, and came to the Czar's attention through his facility with foreign tongues, speaking several Western languages including Latin. He was also a skilled drafter of diplomatic documents and rose to the rank of Vice-Chancellor. He was highly respected by Foreign diplomats.

Ostermann had been born in Germany, the son of a Lutheran Pastor, and was initially employed to translate correspondence between the Czar and foreign courts. As time passed his role grew in importance, and he became responsible for Russian foreign policy, after Peter's death. In 1716, Ostermann was only twenty-nine and junior in rank to Shafirov, who had gone ahead to Merseburg to see Orffyreus' wheel.

Orffyreus was busy arranging to move to Kassel. His machine had been dismantled the previous January, in response to the demand for tax to be paid on his charitable collections. He had subsequently fallen ill, so no wheel existed for the Russian envoys to see. There was little the Russians could do and they had to wait two and half years. In January, 1719, Professor Christian Wolff received a letter from Laurentius Blumentrost, the Czar's new personal physician, Areskine having passed away in 1718. Blumentrost (1692-1755) was born in Moscow, and followed in the footsteps of his father, who had been personal physician to Peter's father, the Czar Alexis. He had attended lectures by Christian Wolff as a student. In addition to his post as personal physician to the Czar, Blumentrost had the additional role of supervisor of that part of the Czar's library concerned with the Cabinet of Curiosities, or Museum.

The letter was prompted by the publication in *Acta Eruditorum* of November, 1718, of news of Orffyreus' machine, and at Peter's request, Blumentrost wrote,

> 'Following the recently published account of the Orffyrean Perpetual Motion in the *Acta Eruditorum* of November 1718, I have decided to approach Your Excellency, and would like to express my high regard for you which I have always had since first I profited from your lectures. There is no doubt that Orffyreus' perpetual motion machine is self-moving because this has been established by the most experienced men of science, and especially by you, who are the most expert in the subject and recognised as such. There is no one better than yourself qualified to judge whether that machine would be capable of exerting its effect when enlarged. I would deem it the greatest favour if you would give me your opinion in this matter, and I would accept your decision. Also I would be pleased to be informed as to whether the machine really is for sale or likely to be, in the future. . .'

Wolff's reply was sent a year later. On the 20th January, 1720, Wolff wrote to Blumentrost,

'It was with great pleasure that I received your letter, but I am in

> your debt for taking so long to reply, as first I had to get information from Kassel. In addition to all that has been published in *Acta Eruditorum*, I can tell you that His Serene Highness the Landgrave has seen the interior structure of the wheel and gave to the inventor a payment of 4000 thalers, together with his word of honour, never to communicate the secret to anyone or to imitate it for himself. Thus, it may be correctly assumed that the wheel is moved by some internal principle. However, I have already shown in my *Mathematical Lexicon*, under the heading 'Perpetuum Mobile' that, up to the present time, we cannot assume that it will have any special advantage because it does not have enough force to do anything. Of course it is possible that the invention could be further improved if it was put into the hands of an expert mathematician. The inventor achieved success through laborious experimentation, but is unable to discover what power lies within it . . .'

Bearing in mind that Wolff has admitted that the machine has an internal power of its own, his recommendation seems rather lukewarm. He wanted to encourage the negotiations with Orffyreus to continue, hoping to reserve for himself the position of 'expert mathematician', so that he, himself would be in a position to 'further improve' the Orffyrean machine. Certainly as author of the article in the *Acta Eruditorum*, he seems optimistic about the machine. There is also the comment that Orffyreus only achieved success through trial and error, and did not understand how his machine worked. It has the same contemptuous ring about it as the comments by Desaguliers, on the subject of Newcomen and his engine, less than a year later — the contempt of a scientist and expert for an uneducated amateur.

Wolff's letter had precisely the desired effect, and Blumentrost, prompted by the Czar, wrote back to the Professor on the 25th November, 1720,

> 'In your last letter, you mentioned that Orffyreus' invention could be of some benefit to the public if it were put into the hands of an expert mathematician to improve it, and this could only be done if the inventor was paid a certain amount of money. His Majesty, the Czar, after due consideration, has graciously advised me that he will spare an unlimited amount of money provided that you, yourself undertake this task and enter into His service. Please write to me if you have any other matters for consideration, and they will be graciously settled by the Czar personally. I can assure you that you will find everything to your complete satisfaction and would there-

fore ask you to let me know what your intentions are concerning this proposition.'

Having received assurance of the Czar's personal interest in the matter, Wolff wrote back to Blumentrost on the 11th January, 1721.,

'It was with pleasure that I received your letter, which states that His Majesty the Czar wishes to take me into His service for the purpose of further improving the Orffyrean perpetual motion machine. At the moment I cannot be certain whether there is anything of value to be obtained from the Orffyrean wheel, but he was given a certificate by the Landgrave stating that there was no fraud within the wheel, so I regard the invention as worthy of being obtained for public use. An expert and mathematician could easily assess how beneficial it could be, but for now I cannot of course, promise that there will be any great advantage to the public.

'Furthermore, before he is paid any money, Orffyreus must agree to several conditions. Firstly, he must allow an examination designed to determine whether his machine is genuine and whether it will be as useful as he claims. The money should be put in escrow during the examination. If His Majesty the Czar wishes to avail himself of my services here, I would gladly receive his orders in this matter, even if I am not officially in His service. I have written to Kassel about this matter (however, without mentioning the Czar) and as soon as I receive a reply I will forward the information contained within it to you.

'Meanwhile although I would like to show my appreciation for the great favour that the Czar has done me, with my gratitude and obedience, I should point out that I do not know whether negotiations with Orffyreus will be possible. Also, will this matter be of sufficient importance to the Czar that he will still require me to enter His service? I am therefore unsure what conditions I should suggest.'

Blumentrost's assistant at the museum was Johann Daniel Schumacher. After at Strasbourg in 1711, he went to Russia and in 1714 became Areskine's secretary. Following Areskine's death, Schumacher became Keeper of the Czar's books and curios and eventually was sent throughout Europe on a mission to buy books, scientific instruments and of course, Orffyreus' perpetual motion machine. On his return to Russia he wrote a detailed account of his journeyings, with the results of his enquiries itemised in an official report, which we will examine shortly. Having stopped in Berlin during his travels, Schumacher wrote to Wolff. It was now the 12th April, 1721.

> 'You should by now have received a letter from His Majesty's personal physician, Mr. Blumentrost, dated 26th February of this year, in which he informed you that I have been graciously ordered by His Majesty the Czar to negotiate with you, not only about the Orffyrean perpetual motion machine but also about His Majesty's further proposals.
>
> 'Having been greatly delayed in my travel by flooded rivers and wretched roads, I was worried that you might, because of the long delay, change your mind and accept some other offer. Because of this, I thought it vital to humbly inform you that I shall leave here after the Easter holiday, and go to Hamburg where I shall have things to order and then directly to Halle. I will have the honour to reveal personally all that His Majesty the Czar has ordered me to do.'

Again over a year after Schumacher's letter, Wolff replied. On the 3rd July, 1722 he wrote to Schumacher, and it was this time only a relatively short distance between Halle and Hamburg.

> 'I am obliged to you for all the trouble you have undertaken on my behalf. I am enclosing my comments about the Orffyrean perpetual motion machine as I promised. As soon as His Majesty the Czar informs me as to what action he intends to take, I will have no hesitation in giving my opinion as to the conditions of Orffyreus' contract.
>
> 'As for the other business, I will make a decision in due course. I would like to know what factors were taken into consideration, in that discussion. It is not that I am unwilling to help fulfil His Majesty the Czar's intentions, in introducing the Sciences into His country and in fact will do everything possible for my part, to bring this about - if only the other conditions will allow me to do so. The biggest problem is the question of whether I should permit myself to move into a rough climate and, risking damage to my health or worse, try to accustom myself to completely different food and drink. In addition, there is also the big question as to whether I would advance my Scientific projects, perhaps only being able to

do that which the Russians will agree upon. I would have to abandon all that I have in mind to do here and no one can continue it without me. I must stress that all this concerns me greatly and hope that you may be able to offer some remedy with respect to other people before I take my final decision.'

One cannot blame the Professor for wishing to be assured of the conditions awaiting him at St. Petersburg. Anyone considering taking up a new position in a foreign country would have done the same. Nevertheless, I think that he never really wanted to move to Russia to work for the Czar. He was perfectly happy where he was, and I think that these comments along with the extended gap between receipt of the letters and his reply were all part of the delaying tactics that he was using. He wanted to be involved in the development of the Orffyrean machine, but he was probably hoping that either Peter would permit him to do his work on it where he was in Halle — or someone else would buy the machine first and he wanted to be on hand in those circumstances to offer his services in a similar role to that being negotiated with Peter.

Wolff then attached his considered opinion of the Orffyrean wheel.

'Impartial Comments on the Orffyrean Perpetuum Mobile, prepared for His Majesty the Czar, Halle, 3rd July, 1722

'**1.**To begin with, it would appear to be beyond doubt that Orffyreus' wheel is not moved by any imaginable external force but rather, its movement is due to the internal weights which are applied in a special manner. My reasons for arriving at this conclusion are:

a) I saw, myself, that the wheel began to rotate with speed and uniformity, without any appreciable external thrust or push until it was slowed from outside. Any attempt at fraud from outside was impossible because the wheel bearings were uncovered on both sides and one could see the axle journals turning in their bearings. Upon request, the wheel was moved from its stand and put on another one.

b) Before translocating the wheel, the Inventor who was performing the test for the officially appointed Commissioners, took out the weights and permitted one of them to be touched, wrapped in a handkerchief. He did not allow the weight to be touched on the end, but lengthwise, it felt cylindrical and not very thick. One could hear the weights landing on the overbalanced side, as though they were swinging, from which one can assume that the overbalancing was caused by their impact. Furthermore there is the testimony of the Landgrave of Hesse-Kassel, who is experienced in evaluating mechanical inventions and had seen the internal mechanism of the wheel and ran it for many weeks in a locked room, keeping the keys himself, having personally locked and sealed the doors and windows with his own seal. He testified both verbally and in an officially printed certificate that the movement of the wheel was caused by nothing more than the weights and that it would run continuously unless the internal structure of the wheel was altered.

2. Since it is impossible, according to mathematical proof, for a machine to run continuously by its own force, some matter from outside must contribute to its motion. That matter cannot be perceived by any of the senses but could be made use of by people who know nature better. I suggest, therefore, that the weights on the wheel's periphery are attached by rods in such a way that when at rest on the lighter side of the wheel, they can be lifted, but when they start to fall, after the wheel has turned, they deliver a force on impact, acquired during the fall, onto a piece of wood which is fixed to the periphery. In this way, the wheel is put into rotation by the impact of the weights, which can be heard. But the force which drives the weights, does not come from the machine itself, rather it comes from some fluid, invisible matter by which the movement of the falling weights becomes faster and faster. Orffyreus' whole invention consists of an artful arrangement of weights, in such a way that they are lifted when at rest and acquire force during their fall, and in my opinion it is this that he keeps secret. This is also consistent with what Orffyreus says, that anyone could easily understand his invention, as soon as he is allowed to look into the wheel.

3. It is possible therefore, that when the internal structure of the wheel has been revealed, some mathematicians may decide that it is not a perpetual motion machine as there is an additional force involved, namely the unknown substance which applies continuous pressure to heavy bodies when they fall, and which adds to the force of their impact.

4. Because this substance is present everywhere and exerts its effects unceasingly, the wheel can be installed anywhere. As a result, it is as if no additional force was involved, and in the end it is the same as if it was a true perpetual motion machine, and many of the maticians will insist that this is so, objecting to this line of reasoning.

5. Until the internal arrangement of the wheel is understood, it will not be possible to discover whether its power can be appreciably increased or not. Perhaps its power can only be increased to a certain degree, after which it will again decrease if someone tries to make it larger. In this case, any attempt to increase its power would be counterproductive.

6. At present, I do not expect any special results from the machine. The one I saw had very thin journals and the axle was hollow; these two features were contributing to the fast rotation of the wheel, and so to its power when the wheel had to lift something. This would not be an acceptable feature in a wheel that was intended for serious development. Such a thin journal would quickly wear out and such a poor axle would soon break down. Besides, there is considerable difference between a machine that is used continuously for work, and one that is run for a moment or a few minutes to pull or lift something.

7. Whoever wishes to buy Orffyreus' machine should decide in his mind whether he wants to have it as a curiosity or for the advantages it offers for moving machinery. In the first case, everything written in notes 2 and 3 should be taken into consideration. In the second case, notes 5 and 6 should be born in mind.'

Wolff's comments make very good sense, and it is amazing how close he came to the secret. We shall see that he had in fact discovered the true principle behind the action of the secret mechanism in the wheel. He did not have an accurate picture of gravity at that time, but he describes it poetically but correctly. It *is* an invisible force; it *is* present everywhere; it *does* apply continuous pressure to heavy objects when they fall; *and* he is correct to say that because of this, the effect demonstrated by the Orffyrean wheel may well be what the inventor *meant* when he claimed that it was perpetual motion, *but* it could, equally, be argued that it wasn't, *because the gravitational force originated from outside the machine.*

The effects of Gravity had been described by Newton, but it was still only vaguely understood. It was believed to be an inherent 'heaviness' associated with all things to a greater of lesser extent. The notion that it might be a force permeating the space outside the 'heavy' objects and applying its influence on them was not generally understood.

Wolff never looked at the Kassel machine, preferring to comment on the lighter Merseburg one. It may be that, as he was satisfied that no fraud could possibly be involved in the Orffyrean machine, he saw no reason to examine the new one. The principle must be the same, only the dimensions had altered. He would therefore, being a thorough scientist only discuss the evidence of his own eyes, ignoring reports by others. He could hardly advise purchase of a machine he had only read about. He found the Merseburg wheel too frail to be of value, but this appears to ignore the fact that it worked. Wolff was a brilliant theorist rather than a practical man and it may be that he was genuinely unable to see how the machine could be improved until he could get his hands on it. Leibniz, on the other hand was intensely practical. He was an expert on the draining of mines and he had designed and built a complex calculating machine, and I think he could see the potential for the Orffyrean machine much more clearly than Wolff. One cannot rule out the possibility that Wolff was well aware of the potential for the wheel, and was playing down the value of the wheel so that he could claim credit for a vast improvement once he had examined it, and specified certain alterations to its internal mechanism.

Schumacher's reply, dated the 12th July, 1722 was written little more than a week after Wolff actually wrote his letter, and indicates a sense of urgency on Schumacher's part, to get the negotiations settled satisfactorily. His letter sent on from Berlin said,

> 'Thank you for your letter of the 3rd July, which included your comments on the Orffyrean perpetual motion machine. I can now officially inform you that His Majesty the Czar intends to found a Society of the learned people in St. Petersburg, who would devote themselves to developing the Arts and Sciences. His Majesty the Czar is well acquainted with your erudition and capacity for work, and he would very much like you to organise and direct the Society, and add to its reputation. I have no doubt that a university will be established later. If you would like to take this post, it only remains for you to send your acceptance. His Majesty the Czar gives the Vice-President 2400 roubles which is 3200 thalers, I am sure He will not give you less. The climate in St. Petersburg is somewhat rugged but very healthy. Fever and infection are seldom encountered; food and drink are as pleasing as that in any restricted market; religion may be exercised in a liberal manner and St. Petersburg is a considerably refined and civilised city and is second to none in Germany. Your honourable wife will also find the pleasant company of ladies who will enjoy the benefits of your presence.'

We can now turn to Schumacher's report, which he wrote for the Czar on his return in 1722, in which he recorded his tasks and their outcome. I have omitted the many tasks required of him that do not concern us here.

> 'In accordance with the orders of His Imperial Highness and Doctor Blumentrost, I the undersigned, was sent to France, Germany, Holland and England in February, 1721. I was authorised and instructed as follows:
> **6.** Contact Professor Wolff in Germany.
> **7.** Talk and negotiate with Orffyreus concerning his perpetual motion machine. If possible, discuss its usefulness with Professor Wolff.
> **9.** Order . . .machines and instruments pertaining to experimental physics from Mr. Muschenbroek.
>
> Herewith I respectfully and humbly present my report at the feet of His Imperial Highness. In it you will discover how I have attempted to carry out my orders and duties. . .
>
> **Item 6.** Nothing has given me as much satisfaction as the positive fulfilment of my commission to persuade the Court Councillor, Professor Wolff to enter into the services of His Imperial Highness, which he eagerly accepted. Initially, I had great trouble because he pointed out that he does not desire more in life. He said he had a gracious monarch, popularity among the students, a steady job and a satisfying income. He could earn a total of 4000 thalers a year made up from writing books, partnerships, discussions and salary. Most important of all, he could live in a moderate climate. Moreover, he is concerned that he might not have sufficient good fortune to please His Imperial Highness; and the nobility and scientists might not tolerate his beliefs; or he might become homesick within a short time; or the damage inflicted to German science by his moving to Russia might prove irreparable.
>
> 'Certainly one cannot blame him for preferring the known to the unknown, however I was able to overcome his doubts and, in his letter of the 22nd September, he affirmed his resolve to enter into the service of His Imperial Highness. His Imperial Highness is well acquainted with the fame and merit which has been well earned by Professor Wolff through his publications, not only on Philosophy, but especially in experimental Physics, and therefore the Professor

has promised to improve, if possible, the perpetual motion machine; and to found a Society of Sciences, keeping it up to such a high standard that the learned people, both foreigners and those of this country shall benefit from it to their satisfaction. Furthermore he has promised to give all his studies and erudition to the service of the people and students. He now awaits only his appointment and contract from His Imperial Highness.

Item 7. Prior to my talks with the inventor of the perpetual motion machine, Orffyreus, I discussed the matter with Herr Professor Wolff. I declared to him Your Imperial Majesty's noble intention, mentioning also that although Your Majesty, for the sake of the common good, would willingly part with a considerable sum of money, you did not want to waste it. Therefore I asked him to give me his honest opinion of it. He replied, 'There is little to be said on this matter, for although Orffyreus has made a wheel, which turns round and pulls a heavy object towards itself, it is not possible to say whether this truly is perpetual motion and will bring great utility to the populace, for it is impossible to see its internal structure'. He instructed me to discuss this with the inventor and report back to him what happened, then he will give his written opinion.

'I set off without delay, straight to Kassel and Weissenstein Castle, hoping to find Orffyreus there, but I discovered that His Serene Highness the Landgrave had sent him, together with the broken parts of his machine, to Karlshafen, which used to be known as Sieburg, about five miles from Kassel, and there a special house was allocated to him so that he could have better conditions to develop his ideas and plans - and also get rid of such a repulsive guest. Your Imperial Majesty may be interested to learn why the inventor had broken his machine.

'The above-mentioned Landgrave had invited Herr Professor 'sGravesande from Leiden to demonstrate to him the physico-mathematical experiments which were published in his book, and they fell into discussion about perpetual motion — whether Orffyreus' wheel really was a true perpetual motion. The Landgrave asserted that it was, and ordered Orffyreus to demonstrate it to Herr 'sGravesande, but without telling him who 'sGravesande was. Orffyreus obeyed the order and demonstrated his machine in the presence of the Landgrave, but 'sGravesande asked so many questions and expressed such a keen interest in learning about its internal structure, that Orffyreus formed the opinion that they wanted to discover his secret, and therefore he refused to show them any more, and as soon as they departed he broke the machine so that there would be nothing to fear.

'After I had spent several days in Sieburg (Karlshafen) with the inventor, pretending to be his very great friend, I revealed to him Your Imperial Majesty's noble intention, and his first question was,

> "does Herr Schumacher have any money?"

'I replied that there was more money available to him than he could imagine, and that Your Imperial Majesty wished to give him high favour and payment for his invention, if it passed the test.

> "Of course it will pass the test, and I am prepared to lose my head if it's not the real thing."

'To which I replied, "let us take two renowned mathematicians, who will first take an oath not to reveal anything about it, then examine your machine and give it a certificate. The agreed amount of money will be put in escrow, and with this we would be satisfied". However I could not obtain his agreement to this. He remained resolute in his opinion that the machine was true, and that no-one could fault it except through ill-will, he could confide in no-one as the entire world was full of evil, malicious and vicious people whom it was impossible to trust. His last words were,
"Herr Schumacher, you put down 100,000 thalers on one side, and I shall put my machine on the other!"

'I could do nothing more, even if I had spent an entire year in negotiations with him, so I returned quickly to Halle and reported to Professor Wolff on what I had done at Herr Orffyreus'.

'It is truly incredible what disputes the perpetual motion has caused. Herr Professor 'sGravesande believes that perpetual motion is not contradictory to established mathematical principles, and although it is impossible to really affirm that Orffyreus' wheel will bring great benefits to the people, he agrees that if it were placed in the hands of skilful mathematicians, it could be brought to perfection.

'Herr Gärtner's perpetual motion machine which I saw in Dresden, consists of a machine in the shape of a grindstone filled with sand

and covered with canvas, which turns backwards and forwards by itself, but according to its inventor could not be made larger. The French and English mathematicians hold all these perpetual motion machines in no esteem, and say it is contrary to mathematical principles. Herr Professor Wolff gave me his opinion of them in writing, which I have thought to incorporate here word for word. It is in line with the opinion of almost all mathematicians.'

Schumacher carried out his commission with exemplary zeal, even meeting Gärtner at Dresden to view his dubious machine. One wonders what turn their conversation took, bearing in mind that Gärtner must have known who Schumacher was and why he was there. Of course Orffyreus had only just left Kassel, and this news may not have become public knowledge yet. This being the case, Gärtner may have thought that Orffyreus was still under the protection of the Landgrave, and that it would be unwise to speak out.

After attaching Wolff's comments, Schumacher finishes this part of his report thus,

'Your Imperial Highness will know from what I have written, that the Orffyrean machine is still far from perfect. Subsequently, Orffyreus wrote to me that a certain Monarch wished to put up all the money asked for, so he intended to reveal his secret to the public once the contract was signed. Because of this, he humbly asked that Your Imperial Highness contribute some part of this recompense too.

Item 9. In Leiden, I bought from Herr Musschenbroek a camera obscura, a lucerne Magique, and some other instruments that are used for injection. These were improved by Herr Professor 'sGravesande. As I mentioned earlier, Professor 'sGravesande published a book *Physices elementa mathematica, experimentis confirmata; Sive, introductio ad philosophiam Newtoniam.* In this book, all machines used for experiments are illustrated, including experiments in mechanics, statics, hydrostatics, physics and optics. Many monarchs order such machines from Herr Musschenbroek, and because I know that Your Imperial Highness is so interested in such machines, I have ordered them too, and I gave to the master 200 gulden as a deposit.'

Petrus van Musschenbroek (1687-1748) worked closely with Professor 'sGravesande, devising experimental instruments, and many of the original designs have stood the test of time. Schumacher's jubilation at securing

Wolff's agreement to enter the Czar's service was premature. After he arrived back in St. Petersburg, his superior, Blumentrost, received a letter from Wolff which indicated that nothing was finalised. The letter which was dated the 24th April, 1723 said,

> '. . . . We must be very careful not to misconstrue the intentions of His Majesty the Czar and must not risk dishonouring him, because the perpetual motion machine was destroyed by the inventor for some reason, as Herr Schumacher knows, and this summer he intends to build a house where one room will be adapted to house a much larger machine than previously. I have in fact written to him, informing him of the noble intentions of His Majesty the Czar, however I have still not received any reply from that strange man. Nevertheless, I shall continue to intercede through the good offices of His Serene Highness's personal physician at Kassel.'

The mystery of Orffyreus' lack of communication was cleared up when Wolff received a letter from the inventor. In a letter to Schumacher dated the 1st June, 1723 he wrote,

> 'At last I can send Orffyreus' answer concerning the perpetual motion machine, together with his first letter that was misdirected and am now awaiting His Majesty the Czar's orders in this matter. I have suggested that nothing should be assigned to the bank at this time and that the agreed amount of money will be paid provided that:
> **1.** The perpetual motion machine is found by me to be correct according to the principles of mathematics.
> **2.** It is found to be efficient for the use for which His Majesty the Czar has designated it.
>
> I can assure you that for both points the interests of His Majesty the Czar will be dutifully observed. If any additional points are found necessary to be taken into account in this contract, I shall expect to receive them in your next letter, together with the agreed purchase price. If the contract is finalised, it should firstly be sent to His Majesty the Czar for ratification, and then, before the contract is executed, it should be sent to me so that I can initial each point, including those introduced by Orffyreus. I await your further orders. . .'

Wolff, when he discovered that Orffyreus had destroyed his wheel, became hesitant to carry out his commitment to work for the Russians. He had already put up several reasons for not accepting the post. All good ones for staying put — except for one thing; the prospect of being responsible for the development of the Orffyrean machine. He may have appeared less than enthusiastic about the Orffyrean machine, but it was he who authored the articles in the *Acta Eruditorum* which recounted the latest information on the invention, and he may have been more enthusiastic than he was willing to admit publicly. While the machine was disassembled he lost his enthusiasm for the Russian post, but following the receipt of Orffyreus' missing letter, it revived.

His first point states that the wheel must conform to the principles of mathematics. In other words, because they believed perpetual motion to be impossible, it could not be defined as a perpetual motion machine, no matter what the inventor said. It is possible that that fact could have affected Orffyreus' claim for a reward for discovering the secret of perpetual motion. It could have played a part in future negotiations designed to force him to reduce his asking price.

The second point states that the wheel must be able to do what the Czar wants it to do. If it any way falls short of Peter's expectations, again a new price could be negotiated. These points were probably included by Wolff for his protection so that if any aspect of the negotiations were found to be displeasing to the Czar, or which seemed to put his own reputation at risk, it would be simple to manipulate the requirements so that the machine did not fulfil them; then all negotiations could halt.

Wolff's brave proposals were irrelevant. Orffyreus would never have agreed to the points outlined by him. Schumacher had already had a taste of negotiations with the inventor and must have known that Orffyreus would never agree to a prior examination of the internal mechanism of his machine. The Professor's next letter, of the 1st June, 1723, included Orffyreus' reply, which had included a copy of the misdirected letter as well as a covering one.

> 'I received Your Honour's letter on the afternoon of the 18th May, (1723) and am bewildered. I cannot understand how or why my reply to you was misdirected or mixed up by mistake. It may have been because I had to post thirteen letters.
>
> 'My reply to your second letter cannot be any different to the first one that was misdirected. In that one I replied as follows:
> **1.** My perpetual motion machine is still not sold and is therefore available for purchase.
> **2.** I am pleased that His Majesty the Czar wishes to purchase it.
> **3.** I am also pleased that you have been assigned to examine and research this machine.

4. The price of the machine is the same as previously: 100,000 thalers.
5. In agreement with your well-intended advice, the price could be reduced by say, 5, 6, 8, or even 10,000 thalers resulting in a final selling price of 90,000 thalers.
6. This price is very low compared to the value of the invention, but should not be felt as more than sufficient for a man such as I.
7. The price can only be reduced by this amount on condition that the money is not to be deposited in another place but right on the spot where the machine is situated, tested, verified, purchased and taken over. This is to prevent possible apprehension for both parties and to avoid any controversy. Unfortunate experiences two years ago have forced me to adhere to such stipulations and not to depart from them, I have learned to be cautious.
8. First of all a proper contract must be concluded, after a preliminary agreement has been reached.
9. Some time ago I expressed the opinion, both verbally and in writing, to Herr Schumacher, His Majesty the Czar's Librarian, with regard to another project. I put everything into my written reply, including the price, and still regard that as binding. I hope to hear news of His Majesty the Czar's decision as to whether to agree to the 90,000 thalers. If the answer is yes, the contract could be quickly concluded and a reliable machine for examination could be quickly constructed with God's help. To date I have nothing more to add to what has already been written.

'If, having given serious consideration to my project, the Czar graciously resolves to have more for less money, then may I at least add this; I have in mind a great 'Treatise on Mechanics' which I plan to publish, with many hundreds of machines and drawings that could be printed at my home.

'In this tract it will be possible to trace my perpetual motion machine through all the stages of its development. As a result, whoever wishes to pay a certain sum of money to receive such a treatise, will be able to select and build at his discretion any machine together with my invention. The Treatise would be in folio in two volumes. It would be ready in about two years from the time I start to work on it. It will be very interesting, useful and beneficial for the public. I would have much more to say about it but, because time is short, I will add only the following, subject to your approval;

'If His Majesty the Czar would advance me 10,000 thalers for this tract, which would increase and promote the publication I could start work on it immediately; in two years he would have quite a large number of these treatises. It would include my perpetual motion machine revealed completely through all stages of its construction, which would eliminate any need for him to contribute anything else. N.B. Sufficient guarantees would be included. In the event that my perpetual motion machine was sold before my treatise was published, my project would still be finished and the treatise published. However , the machine that had been sold would not be included in the treatise, because I want to sell just one kind of machine, not all of them. I have many different kinds of machine all running on different principles. They can be moved by weights, balls, springs, internal gears, internal water, oil, alcohol and wind. I believe it is best that this information is known in advance. Certainly His Majesty the Czar will see the advantages with his enlightened mind more clearly than I could indicate. Please offer my most humble services to His Majesty the Czar.

I remain etc.

'P.S. Carlshafen on Weser, in the greatest haste 11pm, May 21st, as His Serene Highness was leaving my home. He has today, paid a visit for the second time, to our town.'

Orffyreus' reference to a bad experience two years earlier may refer to Schumacher's 'monarch'. In his report to Peter, he mentions that 'some monarch' wished to buy the machine outright. The word 'monarch' is specifically used, rather than 'great prince' such as is usually done. In a short tourist pamphlet, which I acquired, it is suggested that the inventor attracted the attention of Augustus II of Saxony. As Augustus I, he was Elector of Saxony, and as Augustus II, King of Poland. He was a renowned lover of the Arts and Sciences and a most likely candidate for the mysterious 'monarch'. Why the negotiations failed we do not know, but a number of people who attended the tests were also members of the Polish Court and acted as Councillors to the King — as was Gärtner — and it is possible that by this time, negative rumours set in motion by Gärtner or his friends, may have persuaded the negotiators to back down from any commitments they had made, because of doubts about the inventor. To receive Wolff's letter after a gap of two years, must have awakened the inventor's hopes again, and he lost no time in rehearsing his demands, throwing in an additional sweetner to try to tempt the Czar, and then makes the rather cheeky suggestion that the Czar advance him 10,000 thalers.

The inventor mentions that he has just been visited by Karl for the second time. This seemingly irrelevant aside, was added to prove to his would-be purchasers that despite the fact that he had left the Castle Weissenstein, he was still on good personal terms with the Landgrave, and that any rumours to the contrary should be ignored. There is also evidence that Karl's sons visited the inventor more than once, following his exit from Kassel. This does seem to lend credence to the theory that Orffyreus was still in favour with some of the family, if not Karl's mistress, the Marquise de Langallerie.

Orffyreus' undoubted pleasure at the resumption of negotiations with the Czar was short lived. Shortly after Wolff had written in June, 1723, about Orffyreus' lost letter, he received from Blumentrost a letter finalising the Czar's offer to Wolff.

> 'Herr Schumacher has already reported both in writing and in person to His Majesty the Czar when He arrived here, your decision to enter into His service. Associated with this is the possibility of working to improve the perpetual motion machine if an agreement can be reached with the inventor. Also, for promoting the sciences by founding an Academy of Sciences and lecturing the disciplines of mathematics and physics to the benefits of the public, Your Excellency would receive, as Vice-President of the Academy, 1500 roubles a year or special bank thalers, plus moving expenses. The salary will be paid from January of this year. I believe this offer to be both lucrative and proper. In which case, I have no doubt that you will have no hesitation in holding to your decision and will inform me as soon as possible when you will arrive here.
>
> 'With regard to the perpetual motion machine, His Majesty the Czar has decided to put into escrow with the King of Prussia, that amount of money agreed with the inventor. It would then be left to your judgement whether the machine is a real perpetual motion and whether it would benefit the public. In accordance with this, I will negotiate with Orffyreus and inform you of the contract or agreement.'

The inventor's requirements were ignored. Perhaps it was thought that Orffyreus would be forced to compromise if he wanted the sale to go ahead. A position of stalemate was sure to arise soon, when Blumentrost arrived to begin final negotiations, and it would be interesting to know how the deal would have been struck — I suspect that the inventor would have got his way, because Peter wanted the machine and the inventor wanted to sell it to him. In the event all the demands and counter-demands were irrelevant, for what happened next illustrates how precarious everyone's positions were.

On the 8th November, 1723, Frederick William I, King of Prussia, the selfsame man who was to hold Orffyreus' money in escrow on behalf of the Czar, issued a Royal Decree, ordering Wolff to quit Halle and the Prussian dominions, bag and baggage, for evermore, within forty-eight hours, 'bey Strafe des Stranges', on pain of death by strangulation!

Historians have blamed Wolff's expulsion on his detractors. He had obtained his position at Halle through Leibniz's influence, and he urged his students to follow that great philosopher's teachings. They were taught to rely on reason rather than authority, whatever its source, in order to arrive at the truth. Wolff's rivals persuaded the King that the philosopher, with his recommendation to rely on reason rather than authority might, for instance, convince the soldiers to desert. The orthodox theologians regarded such doctrines as heretical and subversive. Such conflicts within the theological faculty affected the entire development of the University, because the majority of students matriculated in Theology. Wolff, with his enlightened, logical approach did not appeal to the orthodox theologists. They wanted him out - and this is the reason given by historians for his sudden forced exit.

This does not explain the suddenness of the order. Forty-eight hours and he was never to return? This suggests something new and unexpected concerning Wolff, that upset the King, to such an extent that he would have had him killed, if he had not fled quickly. The reason for the King's anger seems to me to be obvious. It had nothing to do with the long running murmurs of discontent uttered by the orthodox theologians. Even if they had succeeded in precipitating the King into action, it is unlikely to have warranted death by strangulation, or immediate exile. No — the real reason was the sudden discovery that his most famous and influential Professor was doing a secret deal, behind his back, to enter the service of the Czar of Russia. To add insult to injury, he only found out when Peter, or one of his representatives, opened discussions concerning the holding by the King of Prussia of a large amount of money, in escrow, until his Professor Wolff had decided that the Orffyrean machine was or was not valid.

If the reader still finds this reason insufficient for such an exaggerated reaction by the King, consideration should be given to the nature of King Frederick William of Prussia. Excellent descriptions of the King exist. He is described as brutal and boorish with a hatred of all things artistic or literary. His jester was the ex-Royal Historiographer, Jacob Paul Gundling, an alcoholic who, during his lucid moments had the ear of the King. Gundling's younger brother, Nicolaus Hieronymous Gundling, was also a professor at Halle, and probably the source of much of the gossip concerning Wolff.

Frederick William was frequently in pain from inflammation of the kidneys and gout. On top of this he is said to have suffered from migraines, abscesses, boils, haemorrhoids and terrible pains in the stomach. He was very fat, his eyes bulged and his skin glistened like ivory. He had the most violent rages, quite suddenly lashing out brutally at his children or anyone else near him. These episodes, triggered by small, harmless remarks or even looks, would begin by him swinging his cane, hitting people in the face, sometimes breaking their noses or teeth. To resist or strike back at the enraged monarch was punishable with death. These outbursts are recorded in the dispatches of diplomats at Court as well as in the memoirs of those who were later to help the King's son — the man who was to become Frederick the Great of Prussia - raise Prussia's status within Europe.

These sudden rages of Frederick William do explain the speed and severity of Wolff's punishment. This was a knee-jerk reaction to the discovery that he was being deceived by a senior member of a group of men that he detested. Unfortunately these events which turned the King apoplectic with rage also ended any prospect of Wolff's involvement with the Orffyrean wheel.

Wolff could have entered the Czar's employ immediately, but he thought it better to lie low. To have gone ahead with the arrangements might have been satisfactory at the beginning, but if the Orffyrean wheel turned out to be a failure for what ever reason, he would be stuck for ever in St. Petersburg, unable to return. On the other hand if he rejected the Czar's offer, kept a out of sight, perhaps the King would relent and allow him back. That is in fact exactly what happened. Frederick William reconsidered, although it took him ten years, and asked Wolff back, but it was not until the King died and his son assumed the throne and invited Wolff back as Chancellor of the University that he agreed and returned in triumph.

With Wolff's removal from the scene, the affair quietly disappeared from the pages of the newspapers and took up less and less space in correspondence. Orffyreus, however continued to offer his machine to Peter. Thus in January, 1725, perhaps only a few days before the death of the Czar, a certain Detlev Klefeker in St. Petersburg offered to go to Germany at his own expense to purchase the perpetual motion machine. Correspondence between this man and Orffyreus continued for several years, but without the result wished for by the inventor. Klefeker was Frederick William's representative at the court of the Czar, which is odd, as he appeared to be working for Peter in this particular circumstance. He evidently had a keen interest in the Orffyrean machine. In any event it was all to no avail. Peter the Great died, aged only fifty-two, on the 28th January, 1725.

Thus ended the events surrounding the Czar's efforts to buy the Orffyrean perpetual motion machine. Those whose lives we have examined in relation to Orffyreus' wheel were competent, knowledgeable and discerning, but there is one more who has figured hardly at all in accounts of the legend of Orffyreus' wheel. Although little has been written about this man, lengthy research over a number of years has turned up a certain amount of circumstantial information which may, in the end prove that he was the most reliable witness of all, to the legitimacy of Orffyreus' wheel. His name was John Rowley, Master of Mechanics to King George I, and the wheel made such an impact on him, that he spent the rest of his life searching for the secret of perpetual motion. He may also have been the man behind an attempt to raise one million pounds on the stock exchange, by forming a company set up to buy Orffyreus' wheel.

Chapter Nine

JOHN ROWLEY — MASTER OF MECHANICS

John Rowley was well-known in the eighteenth century, but slipped quickly into obscurity following his death. All, it seemed, that was known about him was that he was Master of Mechanics to the King — and he had been hugely impressed with Orffyreus' wheel.

The only known reference to Rowley and his visit to Kassel, was in Desaguliers' published lectures. However, in the book of *Treasury Warrants* for the 15th May, 1718, a request for payment is made by Rowley,

> '. . . for a large [sun] Dyal engraved and silvered with a pedestal and case presented by His Majesty to the Landgrave of Hesse-Kassel . . . .£64-10-0'

Here is a reference linking Rowley to Hesse-Kassel. Using an appropriate conversion rate (Multiply by 60-80), this equates in modern terms to a sum of between four and five thousand pounds. Clearly the sun dial was a precious work of art, and Rowley had been given the commission to supply the gift by King George I, or one of his ministers, for good reason. Rowley had been appointed Master of Mechanics to the King in 1715.

According to the diary of Mary, Countess Cowper, Lady of the Bedchamber to the Princess of Wales, in February, 1715, she 'spoke to get Mr. Rowley made Master of Mechanicks, which Mr. Bernsdorff promised me', and in that year he is credited Master of mechanics to George I and so continued until the death of his royal patron in 1727. In the book of *Treasury Warrants* it is recorded,

'. . . pay John Rowley, instrument maker, £75 for the half year from 1714 xmas on a yearly sum of £150, which the King is resolved to grant him in consideration of the good proofs of his great skill and knowledge in the mechanic arts by several inventions which he hath shown the King and as an encouragement to him to proceed to the improvement of those arts which are or may be of great use to the Office of Ordinance as well as a public benefit. . .'

£150 as a salary was equivalent to around £10,000 today, and although not a great sum, it must have been a welcome addition to Rowley's income from his shop. On top of this, he earned money from various items, such as the sun dial mentioned above, that he made for the King. His job was not just a sinecure — he got it because of his skill 'in the mechanic art'.

Rowley was born in Lichfield, the son of a long cutler, or sword maker, called William Rowley. He was the youngest child, and his date of birth is unknown, but he travelled to London to become apprenticed to Joseph Howe, a mathematical instrument maker. His apprenticeship indenture is still in existence, and it is dated 23rd November, 1682. When a boy was apprenticed to his Master, he was officially bound at the age of fourteen, which puts Rowley's birth date at sometime in or near 1668. Freedom from apprenticeship was usually granted at the age of twenty-one, and records show that Rowley was released as a freeman on the twenty-sixth February, 1690, so that was probably the date of his twenty-first birthday, so he was born on that date in 1669.

An Analemmatic Dial bearing his name is dated 1698, and one can assume that he was in business for himself by that date. From then on the output of instruments bearing Rowley's name was remarkable, not just for their excellence of workmanship, but also their extraordinary variety. Many of these pieces are still in existence, and may be seen at the Oxford Museum of the History of Science. Instruments bearing his name include Drawing instruments, Pocket dials, Proportional Compasses, Parallel Rulers, Protractors, Artillery Scales, Dialling Spheres, Gunther's Quadrants, Globes, Planetaria (both Ptolemaic and Copernican), Levels, Sextants, the 36 inch, the Standard of Length and of course the celebrated Orreries.

In addition to these, one must include numerous sun dials, telescopes, calculating machines, pocket watches and various surveying instruments. By 1704 Rowley was regarded as the foremost mathematical instrument maker, because in that year, John Harris, Fellow of the Royal Society writes in his *Lexicon Technicum*,

'So I have designedly done Justice to such Ingenious and Industri-

> ous Articifers, as do truly deserve the greatest Encouragement for their skill and accuracy in the making of these fine Instruments: Such are Mr. John Rowley, Mathematical Instrument Maker at the Globe under St. Dunstan's Church in Fleet Street.'

He continues to praise Rowley in his second volume which appeared in 1710. Similarly, words of praise about Rowley's workmanship appeared from time to time in the daily papers, particularly that of Sir Richard Steele. Steele had ascribed the invention of the Orrery, an extraordinary device which mechanically demonstrated the movements of the earth, planets, sun and moon, to Rowley, but this was incorrect. The Orrery was originally produced by George Graham F.R.S., (1673-1751) and Thomas Tompion who made a planetarium, demonstrating the heliocentric system in or near 1706, for Prince Eugene of Savoy, who fought under the command of John Churchill, the Duke of Marlborough, at many battles. It seems that the 'Orrery' was in Rowley's workshop awaiting shipment to the Prince, along with some of Rowley's own instruments. Lord Orrery saw the device and asked Rowley to make him one as well, so Rowley copied it and subsequently made several others, containing additions and improvements of his own. According to our old friend Desaguliers;

> 'Sir Richard Steele, who knew nothing of Mr. Graham's Machine, in one of his Lucubrations, thinking to do justice to the first Encourager, as well as the Inventor, of such a curious Instrument, call'd it an Orrery, and gave to Mr. J. Rowley the Praise due to Mr. Graham.'

Another writer of the day, Earnest Budgell, marvelled that,

> 'The Instrument which was invented by him, and bears his Name, is an undeniable Proof of his mechanick Genius. There are so many different Motions in his Machine, that, I have heard his Lordship say, it had almost turned the Head of the ingenious Artificer, whom he employed to make it'.

Desaguliers' version is correct, but the main point is not so much who designed it, but who made it. John Rowley made this most complicated piece of machinery, which is still in existence and is, I believe in the Science Museum in London. It was sold in 1974, by the family of Cork and Orrery for £29,000. In 1937 the instrument, which needed repair and cleaning, was found to consist of a brass radial frame, driven round a fixed central stem by means of a worm and large worm-wheel, and carrying several trains of wheelwork by means of which the relative motions of the three bodies, the earth, sun and moon, are produced. The Orrery was found to contain no less than fifty-six tiny gear-wheels, each handcrafted.

Rowley was commissioned to supply instruments for the proposed Observatory at Trinity College, Cambridge in 1703, and in the process became acquainted with Sir Isaac Newton, for whom he made several instruments. A sun dial, at Cranbury Park, near Southampton is credited to Newton and he is known to have spent the last few years of his life at Cranbury with his niece. Newton is said to have erected the sundial in 1720, but an examination of the dial reveals that it is engraved with the words, 'John Rowley London fecit'. Probably Newton designed it in consultation with Rowley, who then constructed it and, more than likely erected it on the site, to Newton's directions. I suggest that Rowley erected it for two reasons; Newton was by then seventy-seven years old and unlikely to have been physically able to site the sundial as it is mounted on a hefty stone pedestal, and there is considerable evidence that Rowley liked to erect all his own sundials. It was for this reason that he went to Kassel.

In 1710 two brothers arrived in London on a tour of the sights. Zacharias Conrad von Uffenbach and his brother Johann Friedrich, spent five months in England, and together visited old and new buildings, libraries, virtuosi collections, theatres, popular amusements such as cock-fights and bear -baitings, and the shops of instrument-makers. Eventually they arrived on the 28th October, 1710, at John Rowley's establishment. There they were shown various items;

> 'There were no other instruments completed at the moment, which was a great pity as he is considered to be one of the best mechanicians in England'.

The brothers confirm that in 1710, Rowley was still regarded as the leading instrument maker in England. The word mechanician was popularly used at the time, and denoted a person skilled in the science of mechanics rather than an operator or maintainer of mechanical devices.

Sir Isaac Newton had a particular interest in the Royal Observatory. The observations made by the Royal Astronomer, John Flamsteed (1646-1719), were of vital importance to Newton's theory of the motion of the moon. Much bitterness existed between the two men over the fact that Newton had disclosed those observations to others, and the ill-feeling was exacerbated when Newton decided to have all of Flamsteed's instruments, largely purchased by him, checked and examined by a special Board of Visitors set up for the purpose. Rowley was appointed to carry out the inspection and repair or replace any instrument found defective.

Rowley had also been commissioned to make 'an engine for pumping water' out of the Thames, which was operated by the flow of the tide. Turning to the Treasury Books we find an invoice as follows,

> 'John Rowley, master of mechanicks, for making a dam before and behind the engine, for clearing the old foundation, for setting down a new frame, 26 foot long and 11 foot high, broad enough for the twelve foot wheel; for the new wheel of twenty-four foot diameter and twelve foot broad; for the new brass engine with brasses to the crank, forcing rods and a new crank etc . . . £740.

Following the successful commissioning of the pumping engine, which was to supply water to the Royal family at Windsor, Rowley was requested to become engine keeper and to maintain the device in working order. He subsequently became engine-keeper to the Board of the King's Works and looked after all similar engines within that organisation. His superior was the surveyor of the King's works, Sir John Vanbrugh.

Vanbrugh is known for his plays and the fact that he designed Blenheim Palace, near Oxford for John Churchill, the Duke of Marlborough, and his wife, Sarah. Vanbrugh designed the palace, and the King's works constructed it. The craftsmen working at Blenheim were the top men of the day in their chosen field of expertise. One of these craftsmen was John Rowley. He was commissioned to supply four sundials to be erected at Blenheim Palace, on pedestals, just as he was to do later at Cranbury Park for Newton. Three of the sundials survive. Rowley charged £60 for each of three of the sundials and fifty for the fourth one, plus £5 for the 'journey to Woodstock to Sett up the Dialls and taking several levells'. Rowley was a painstaking craftsman, and would not have allowed anyone to finish a job that was his. Any instrument he made he always installed personally and this is why he was at Hesse-Kassel — to erect accurately the sundial that he had made for Karl.

These craftsmen were generally regarded as the leading virtuosi of the day, each within his particular field, and of course they had worked together for many years on different projects. Men such as Grinling Gibbons, whose skill at woodcarving was unequalled; Sir James Thornhill portrait-painter of the rich and famous; Langley Bradley foremost clock-maker who designed and built St. Pauls's clock and others at several stately homes, and John Rowley. All these men belonged to a little-known organisation or club, called *The*

Society of Virtuosi of St. Luke. There is in existence at the British Museum a brief account of the Club and its transactions, plus membership lists and included is the name of John Rowley.

Desaguliers had mentioned that Rowley 'talked so much of the wheel he had seen at Hesse-Kassel', and it may have been to some of these influential members of the Club of St. Luke's, that he spoke. The Club had been formed by Sir Anthony Vandyke, the painter, in the previous century, and, apart from a break during the civil war, it had continued, revived at the restoration of the monarchy by Sir Peter Lely. It was originally intended as a meeting of painters, sculptors and art lovers, but in the eighteenth century the scope of the members was broadened to include all those leading craftsmen in various fields of the arts. Rowley was invited to join for his painstaking, delicate technique in the field of mechanics, an unusual inclusion, but indicative of the high regard in which his fellow-members held him.

Each year a steward was chosen, and on St. Luke's day or near, he had to provide a dinner for the rest of the club and also entertainment. We are not told what the entertainment consisted of, but the names of the various different venues are well-known as some of the most famous — and notorious — taverns of the time. The Steward for the year 1720 was John Rowley, and the list of members provided in the manuscript at the British Museum, was taken from John Rowley's pocket book. They include, Michael Dahl, portrait-painter; Gibbons the wood-carver and sculptor; Robert Child son of the founder of Child's bank; Hugh Howard, portrait-painter and art collector; John Chicheley, founder-member of the Society of Antiquaries; Thomas Walker, art collector and said to have been immensely rich; Christopher Wren, son of the great architect; James Wootton, painter of dogs and horses of the rich; James Gibbs, architect; Sir James Thornhill, portrait-painter; and other influential men of the time. These were the men Rowley discussed his visit to Kassel with. Other people who would have been on the receiving end of his enthusiastic praise of the Orffyrean perpetual motion machine would have been those with whom he mixed, on a daily basis, around and about the area of Fleet Street where he ran his business, including Lord Orrery and perhaps Rowley's contacts at Court. Desaguliers actually mentions the gossip about Orffyreus' machine emanated from Rowley and says that he was so convincing, that a great many 'ingenious' people were persuaded of its truth. 'Ingenious' people might well be a reference to the membership of St. Luke's Club and knowing Desaguliers' attitude to uneducated people, it is unlikely he would have taken any notice of anyone else who challenged him to answer the claims that Orffyreus' wheel was genuine.

Desaguliers states that it was 'About the year 1720 and 1721' that Rowley began to talk of the wheel. This is more precise than the wording might imply, because Desaguliers could refer to his account of the episode in the *Philosophical Transactions* for the year 1721, because his lecture on the subject is dated August of that year. And it was partly in response to the Rowley's excitement that Desaguliers gave his lecture on the impossibility of perpetual motion.

The presentation of the gift of a sundial to Karl, would have necessitated Rowley's presence, and because he kept it in his possession to look after it, would have brought it to the presentation ceremony himself. The presentation would require the presence of the King, as it was a gift from him to Karl. Rowley would have travelled with the King's entourage, both to and from Hanover. In 1720, George travelled to Hanover. Negotiations to end the Great Northern War were proceeding slowly but promisingly. That summer was remembered at Hanover as a particularly festive one. The King was in buoyant mood and acting as host to King Frederick of Sweden, Karl's son, plus an estimated additional seven-hundred guests.

The gift George presented to Karl was in recognition for his services to Britain, fighting at the head of his own forces alongside Marlborough and Eugene, prince of Savoy, and for the mercenaries supplied to Britain and her allies by him. A suitable moment to present the gift was required and I think that it could have been the occasion of his birthday, 14th August. Linked to this was the fact that negotiations to end the war were drawing to a close, and it was customary to give gifts to all those involved in the negotiations. A letter sent to the *Daily Post* from Hanover, reported on the 9th August that ' King George and the King of Prussia are here and the Landgrave is expected,' which fits in well with Karl's birthday.

Rowley entertained his fellow members of St. Luke's club on the 6th December and it was then that he told them of his memorable visit to Kassel. The official date of St. Luke's day was the 18th October, but rarely did the club hold their feast on that day so it was not unusual for Rowley to arrange his Feast for the 6th December, 1720. He was probably prevented from doing so earlier as King George did not return until the 12th November, and it is unlikely that Rowley would have travelled on his own, preferring to make the journey in the company of the rest of the entourage, arriving back in England around the same time.

Between the occasion of the presentation to Karl, and his return to England, Rowley made the trip to Kassel to erect the sun dial. Possibly Rowley had already mentioned that he was interested in Orffyreus' perpetual motion machine. In which case Karl would approve of Rowley's interest because of his position as Master of mechanics to the King, and the possibility that he might be able to influence George, directly or indirectly, to purchase the machine. King George's minister in residence at Kassel, James Haldane, who was present at the celebrations at Hanover, could have travelled with Rowley to Kassel, as interpreter.

Rowley discussed his visit to Kassel with many people, but there seems no evidence of his further interest in perpetual motion, other than the remark by Samuel Johnson, that Rowley spent the rest of his life investigating the subject. There is circumstantial evidence of a sort, that Rowley did do some research into the subject. During the tour of the Russian Librarian, Daniel Schumacher, a visit was made to the library of Robert Harley, Earl of Oxford. The keeper of the library was Humfrey Wanley (1672-1726), a keen student of old books and documents. During his time at the Harleian Library he kept a journal in which he recorded the names of all the visitors to the library. He noted several visits by members of the St. Luke's Club, and it is not surprising to learn that John Rowley also came to see the books at Harley's library. His first visit took place on the 31st January, 1721, and Wanley records that 'Mr. Rowley came and saw some of my Lord's fine books'. The second visit came scarcely two weeks later on the 13th February. The third visit occurred on the 7th April, 1721, seven weeks later.

The reason for including this information about Rowley's visits to the library is that they only began after his return from Kassel, and then he made three visits in a short space of time. There were books on the very subject that Rowley wished to know about. One of the books in the Harleian Library was the Marquis of Worcester's *Century of Inventions* (1655), which included his description of a perpetual motion machine which he is said to have exhibited before King Charles I. Another book in Harley's library, was written by Thomas Tymme, a Professor of Divinity. It consists of an account of the invention by Cornelius Drebble (1572-1634), engineer to King James I, which I described earlier. A third book might have been *Mathematical Magic* (1648) by Bishop John Wilkins in which the author describes attempts at perpetual motion machines. Many other books in the library contained references to perpetual motion machines. These books could have been consulted by Rowley on his return from Kassel, and it was, I suggest, for the express purpose of trying to discover Orffyreus' secret that he went to the Harleian library.

John Rowley never discovered how Orffyreus achieved his perpetual motion machine and of course he had a business to run. In the *Calendar of Treasury Papers* for June, 1723 is recorded,

"A petition from John Rowley ... Pursuant to His Majesty's orders, 'petitioner has lately repaired the great solar system & made several new additions to the same, particularly a new pedestal & sphere richly carved, gilt, glazed, &.,' for which he conceives that he deserves £500; has likewise made for His Majesty's own use a curious silver universal pocket dial, value 30 guineas, and there is due to him upon a late allowance of £100 per annum, £150 by sign manual, dated 24th June, 1723: prays payment."

Rowley is here demanding the equivalent of between forty and fifty thousand pounds! There is little doubt that he was a wealthy man. The above payment request shows that Rowley was still in his prime and capable of turning out works of great skill and craftsmanship, and aged fifty-four. The Master of Mechanics was probably as knowledgeable as anyone about ways of driving a wheel the size of Orffyreus'. He had experience in designing and constructing an enormous variety of machinery. At one end of the scale he built a huge tide-driven water wheel for Windsor Castle; he constructed in concert with Desaguliers a new ventilation system for the Houses of Parliament; he built telescopes for Newton; astronomical instruments and sextants for the Navy and the Royal Observatory; incredibly accurate scales for the Royal Artillery; and at the other end he built pocket watches; compasses; clocks; calculating machines and of course the Orrery.

One Orrery which he made for the East India Company was said to contain 'above eighty different wheels . . . One single turn of the instrument sets all the several bodies, the sun, moon and earth and all the planets, in their several and distinct motions. . . . the eclipses and phases of the planets and all the different regards of each to one another appear plainly and evidently . . . The author of it is Mr. John Rowley, who himself was there to show it to us and explain it to us.' Rowley also took it to pieces in front 'of His Majesty King George I and to his great satisfaction showed him every part of the contrivance'.

Rowley did not just make the machines to some other person's instructions — he fully understood what he had made, how it was designed to work and what it demonstrated. He had a natural ability and interest in how things worked and he would at once have understood the limitations that Orffyreus had imposed on himself by constructing so large a machine. It would have been apparent to Rowley, that no secret mechanism could possibly drive such a large construction through contact with the bearings. The only contact point with sufficient purchase to drive the machine at its full speed of twenty-six revolutions a minute, would have been on or near the circumference of the wheel, and such a mechanism would have been instantly obvious to everyone. Rowley would have inspected the bearings and supports for signs of deception, because it is certain that he had heard all the rumours about Gärtner's hidden servant in an adjoining room, and surely eyes so searching and experienced as Rowley's would have known what to look for if fraud was involved.

Rowley continued as Master of Mechanics until the King died in June 1727, when he retired. He was succeeded in his post as Master of Mechanics by Dr. Robert Smith, F.R.S., Master of Trinity, second Plumian Professor and Director of Cambridge Observatory. That Rowley raised the status of Master of mechanics to new heights, is evidenced by the rank of his successor. His other role as Engine Keeper to the Board of Works he retained until his death on the 14th January, 1728. He was buried in the Old Vault at St. Dunstan's in the West, right by the site of his first shop.

If Rowley could have persuaded someone to take notice of his opinion on the Orffyrean perpetual motion machine, he could have changed the course of history. Unfortunately the South Sea Bubble occupied everyone's minds. Hundreds of people lost their life savings, potential sponsors or investors were scared off. Despite the advertisement by persons unknown asking one million pounds for the setting up of a company to purchase the secret of perpetual motion, no response was recorded. No-one was going to risk money on an invention that appeared to have been superseded by the steam engine. The story did not completely die. One man did try to revive interest in the Orffyrean machine. Indeed he commissioned a new perpetual motion machine from Orffyreus in 1745, when the inventor was in the last year of his life. However before that, Orffyreus was to undergo arrest, imprisonment and enduring disgrace.

Chapter Ten

THE ARREST

Orffyreus was now living in Karlshafen in a house given to him, by Karl, and he was also advanced five years salary. It is an indicator of the Landgrave's regret at having to accede to demands for the inventor's removal. By granting him five years salary, he was guaranteeing Orffyreus an income and a job for the next five years which must, he believed, give the inventor time to reap the rewards of his invention and provide time to get his new venture off the ground.

Professor 'sGravesande, meanwhile must have wished he had never got involved in the Orffyrean project. After his examination of the machine in 1721 he had tried to calculate mathematically, the reasons for its action. Like Professor Wolff before him, he had decided that the best avenue of exploration lay in the study of the impact of the weights. Again as Wolff had done, he assumed that the sound of the falling weights was a clue to the motive force behind the wheel's power. He was sure that the impact of the weights on the circumference of the wheel gave impetus to it.

Having pondered on the problem for some months, he issued a small booklet entitled *Some Remarks concerning perpetual motion,* in which he defended his reputation, which had been somewhat soured by the publication of his letter to Newton in August, 1721. He theorised on the reasons why Orffyreus' machine worked and came very close to the truth. In part III of his argument he made the following remark,

> '. . . .after my letter to Mr. Newton, I changed my opinion concerning the nature of the Force on which impact depends, and which I no longer believe can demonstrate perpetual motion, and which appear to me to be incontestable consequences of a generally

> recognised principle. *Nevertheless, I cannot be persuaded that it will never be possible to demonstrate that it is contradictory to construct a machine which will have its own internal mechanism for increasing its power as a consequence of the laws of nature.* [my italics] We have too little knowledge of these laws, and there is little prospect of them ever being sufficiently revealed to arrive at a similar conclusion. It seems to me, on the contrary, from what we know of these laws we should try to consider the very possibility of such a machine which we have described, even though the human Art may never attain it.'

Jean Bernoulli (1667-1748) responded to 'sGravesande's publication. He was a member of a famous family of Swiss mathematicians and was, himself, a major contributor to mathematics. He was a firm supporter of Leibniz in the controversy with Newton over who first discovered the calculus. During his final years he worked mainly in the field of mechanics. In a letter to 'sGravesande he wrote,

> 'I move on now to your comments on the possibility of perpetual motion, which were written on the occasion of your examination of the effects of the machine at Kassel. Several years ago, marvellous things were written to me from Germany about this machine, and even a view of the exterior was sent to me. Now, I have been assured that the secret was communicated to His Serene Highness, the Landgrave of Hesse, under an oath of silence, and he was allowed to examine the internal structure of the wheel. Afterwards, his Serene Highness was quoted as saying to his ministers, that he believed the machine to be a true perpetual motion machine, and in addition, it was so simple and easy to construct that he was amazed that no one had managed to invent a similar machine before Herr Orffyreus.
>
> 'As for myself, I do not know what to think about it. It seems to me that a purely artificial perpetual motion is impossible. I base my judgement on the general law of Statics. According to this law, the common centre of Gravity of all moving parts in such a machine must descend continuously unless it is wound up, as is practised in clocks and other automata. I note your opinion is to the contrary because, as you point out, we do not know the laws of nature sufficiently to demonstrate the impossibility of perpetual motion. But is it necessary to know all the laws? Is it not sufficient if only

> one law clearly indicates to me that it is not contradictory to assume that such a thing really is impossible? Whatever other laws there are that are unknown to me, it is certain that the laws of nature cannot contradict or negate each other.'

Bernoulli goes on to say that any motion which exists in nature can be used to support a perpetual motion, as suggested by 'sGravesande and, 'I can assure you that a perpetual motion either created, or just supported by an external movement which exists in nature is not absolutely impossible. On the contrary, it is apparent in the motion of water, rivers and the sea, magnetic material and in numerous other similar instances. However, I must remind you of the definition of perpetual motion which you gave on page four. You say that, "in Mechanics perpetual motion is defined as a machine in which movement is independent of any exterior agent, and the movement will never stop by itself as long as the materials [of which it is made] endure". I must ask you, therefore, why the use of such natural phenomena to drive a machine cannot be regarded as an exterior source of power? In these instances such machines cannot be regarded as purely artificial perpetual motion machines, but rather, as a combined perpetual motion because their motion is assisted by nature. I am convinced that the machine at Kassel is of this type. Perhaps there are some magnets or springs inside that maintain the initial push given to the wheel. I even think that I have discovered a way to make such a machine. I only wish that some capable mechanician could execute my project. I was informed that Herr Orffyreus is asking a reward of 100,000 thalers for revealing the secret of his machine, I would settle for far less.'

Professor Allemande, 'sGravesande's biographer and editor, remarked that:-

> 'The strongest objection that can be made against Orffyreus' wheel is that perpetual motion is impossible. Most scientists are in agreement on this, and to assert to the contrary is to risk inviting ridicule, and would give one's ability in mechanics a bad reputation. Such people are considered dreamers or like those seeking the philosopher's stone in chemistry'.

This is the crux of the problem. Proponents of perpetual motion have always been called dreamers. No one dare publicly discuss such possibilities, even if it can be shown that no accepted scientific principles are being over-turned or over-looked.

The history of Orffyreus is less well documented from this time on, although there is one infamous period for which records still exist and which we shall examine shortly. The planned super-wheel at Karlshafen never happened. The inventor's salary ceased to be paid in 1730, with Karl's death and from then on the inventor had to obtain an income earning post. According Strieder, Orffyreus' his brother, Gottfried had travelled to England, his wife had died and his maid was no longer with him. The catalyst that set these things in motion will be discovered shortly, but in the mean time, under these circumstances it would have been very difficult for Orffyreus to undertake the construction of so large a machine as he had planned, owing to the loss of his trusted assistant. In addition, he must have been left with the responsibility of looking after his young children himself. Although we know he had four, we do not know whether some or all of them were with his first wife. In any case he had at least one daughter to look after, and despite his sudden comparative wealth, no new wheel was possible for the time being.

Following swiftly on Orffyreus' move to Karlshafen, Gärtner decided that the inventor's move from Kassel indicated that he was no longer under the protection of the Landgrave, and therefore fair game for one of his forays into print. His publication appeared in 1723, and it is obvious that he must have been harbouring a desire for revenge against Orffyreus ever since the inventor had come under the protection of the Landgrave. The Court Master-Modeller's document was another wager — a re-hash of a previous publication, and cannot have enhanced his reputation.

On the 28th November, 1727, Orffyreus' maid signed a statement which sounded the death-knell for his wonderful machine. An account of the statement is given in Strieder's biography, and that is the one which I have reproduced here.

> 'Orffyreus' maid, Anne Rosine Mauersbergerin, born in Drebach near Annaberg in Saxony, 38 years of age, stated that she had served him for many years, since 1711, and knew the full facts and circumstances surrounding the Orffyrean wheel. It had been turned manually from its very beginnings in Gera and had never moved by itself; even in places such as Merseburg and Weissenstein when it had been locked, Orffyreus, his wife, his brother Gottfried and her, the maid, had taken turns to rotate the machine. He had promised her 2 Groschen for each hour of turning, and she had received 9 Reichsthalers as salary only a few months ago. Gottfried, who had done most of the turning, had received a hundred Reichsthalers. She was made to turn the wheel by night and day, since it was feared that someone might look to see if it was still rotating at night. When

the archimedean screw had been attached to it as well, it had become very heavy to turn. The posts had been hollowed out and contained a long and thin piece of iron with a barb at the bottom which — according to the statement — was attached to the shaft journal. Turning was carried out from Orffyreus' bedroom which was close to the machine, on a shelf behind the bed. If they would take her to the place and the room in Weissenstein, she would be able to show how everything had been arranged. The strip of iron had been forged by the Court Blacksmith in Kassel. 'She, the maid, Gottfried, his brother, Orffyreus' wife and daughter . . . had all been forced to swear a terrible oath which she had kept in the original drawn up by his own hand and of which copies are attached.

'When the wheel had to be turned by day and night for eight weeks in Weissenstein, she once complained that she could not carry on like this, particularly as it happened in the winter's cold. He replied that she should not fret and promised to pay her so that she would be content. On being asked how he could have instructed people turning the machine to stop it or start it up again or make it change its direction of rotation when he, himself, was in another room and in the presence of strangers who had come to see the wheel, she replied that she had recognised his wishes either from his manually starting the wheel or from his coughing, clearing his throat, spitting or a similar sign. He had initially refused to pay her the outstanding money due to her for turning the wheel, but when she threatened to lodge a formal complaint in Kassel and tell all, he finally consented to pay her'.

Strieder follows this very damning statement with the remark that he has copies of the statements of six other witnesses who were all under oath to Orffyreus. He does not reproduce them, merely commenting that they all agree on the main matter — that the whole thing had been a swindle. Leaving aside for the moment the maid's testimony, two things occurred to me; firstly, the fact that Strieder does not produce the other statements could imply that they might weaken his case in some small way. Presumably he quotes the most damning statement, but he does say that they agree on the main fact, that there had been a swindle. One could also infer from this that other facts within those statements do *not* agree. But this is only speculation. The maid says she recognised what Orffyreus wanted her to do when turning the wheel, by listening to him coughing, clearing his throat or spitting. I have not tested the theory, but it seems to me that the sound of someone spitting in one room, within a castle, might be rather difficult to hear from an adjoining room. Castle walls tend to be very thick and therefore difficult for sound to penetrate. There might of course have been some kind of hole which allowed the sound to travel into the next room.

The maid's evidence is hard to refute. The only way we can judge it is to study it and apply common sense, ignoring the emotive contents which accuse a man of cheating. As I said at the beginning of this book, one fact alone should cause doubt to begin to surface, and that fact is this. The maid claimed that a piece of iron secreted within the hollowed out posts supporting the wheel, acted on the journals of the axle. The wheel was twelve feet across, and eighteen inches thick. I have conservatively estimated its weight to be at least seven hundred pounds, and my reasons for arriving at that figure will be explained later. This mighty construction rested on two bearings of only three-quarters of an inch diameter. First of all it would not be possible to maintain a speed of any kind driving a wheel of this size through the bearings. It would be even more difficult to accelerate the wheel by this method; and to suggest that one could drive the machine *and* turn an archimedean screw or raise a box of bricks weighing seventy pounds is ludicrous. And — how could you effect the connection? The bearings were always left uncovered for inspection, so the only point of contact invisible to the onlookers would be *under* the bearing. The supporting posts were of necessity vertical; how wide would the post have to be to incorporate a long thin piece of barbed iron which had to act *horizontally across* the *underside* of the bearings? How could you apply force to the underside of the bearing with a piece of barbed iron? What kind of shearing force would the barb have to be able to resist — and do this for eight weeks? If you can achieve this, there is another problem. Translocation of the wheel meant that it was lifted from one set of bearings and replaced on a second set. Signs of the barbed iron would have been visible to all the many people who inspected it. The maid says the inventor gave her secret signals to indicate what was required — surely none were required — would it not have been obvious since the machine had to be started with a gentle push in one direction or another?

The first things that all the examiners wanted to inspect were the bearings. There are numerous occasions when it is mentioned that the bearings were left uncovered in order that the inspection of the wheel might continue without hindrance. Descriptions of the bearings abound. Professor 'sGravesande described them as being 'about three-quarters of an inch in diameter' and 'I have examined these bearings and am firmly persuaded that nothing from without the wheel in the least contributes to its motion'. I must point out the obvious here. Many people, probably hundreds, examined the machine, and it seems to me that we insult their intelligence to suggest that despite all their care and determination to arrive at the truth, not one of them, over a period of fifteen years ever found signs of a secret connection to the bearings; or put two and two together when Orffyreus coughed and spat, to conclude that he was giving signals to an accomplice.

There is also in the maid's testimony the claim that they had to turn the wheel day and night. During the Weissenstein test, it ran for fifty-four days. In his account of the maid's statement, Strieder says that the maid was asked how Orffyreus made his wheel-turners aware of what he wanted them to do. This is an important clue to the so-called statement. What we have here is not really a statement at all. It is a record of a series of questions and answers — an examination. The maid has accused Orffyreus of fraud. Many questions are put to the maid to try to elicit the truth. What appears to be volunteered information is actually a response to specific questions. For instance the questioner asked how was the wheel able to turn for eight weeks non-stop. Obviously everyone involved in the fraud had to take their share of turning. Even the inventor's wife was included — she had been dead for a year, so there was no danger of her denying anything! What if someone had visited the wheel at night to see if it was turning? The maid's reply would be as we saw above, that she and three others were made to turn it day and night. The Magistrate might ask, 'wasn't it very cold at night? ' The maid claimed that Orffyreus would pay her well. Another question might have been, 'how did you know what Orffyreus wanted you to do with the wheel? 'She would tell him about the pre-arranged signals.

If, as she claimed, she had to share the turning of the wheel with Orffyreus, his wife, and his brother, then each had to turn the wheel for an average of six hours in every twenty-four for two months! Orffyreus would have to be on hand most of the time during the day, so if he took his turn then he must have done it at night — when did he sleep? The maid says that Gottfried did most of the turning, so let us assume that he did it for eight hours a day, that still leaves sixteen hours shared among the other three, say five hours and twenty minutes each. I am sure the reader sees the point of this, I do not believe that anyone would be capable of turning that machine for say, three hours at a time twice a day for two months — and that is assuming that the wheel *could* be turned. We have ignored the fact that the wife and the maid would be also bringing up one or more children; washing, cooking sewing, socialising, etc,.

The maid says that all of them had to swear a terrible oath. We have had a taste of Orffyreus' oaths, he swore one in his *Apologia* concerning Wagner, and called down on himself all manner of evils if he was lying. It is not surprising that the inventor should require his family to swear an oath of secrecy. Their future depended on no one discovering the secret; he had to urge members of the family who might be privy to it to promise not to tell anyone. Swearing an oath of secrecy might seem to be going too far, but taking Orffyreus' nature into account and the times he lived in it is probably understandable. He was suspicious of everyone. He was very religious and believed that everyone would be honour-bound to keep a secret if sworn on the Bible and in the most awe-inspiring conditions. The oath we are about to encounter would certainly impress anyone today, and when you consider how much more God-fearing people were in his day, it is easy to see why Orffyreus made them swear.

The maid says he even made his daughter swear this terrible oath. Orffyreus was married in 1712, so in 1716 when they were doing the eight week test, she could not have been more than three years of age! In his *Apologia* the inventor comments that although they were very happy, they were not blessed with children immediately. So even if we only delay the arrival of the first child by only one year, she could not have been more than two years old. Perhaps the maid meant that she swore the oath in 1724, at the time of her own oath. If so, she would still only have been about nine or ten.

The maid's oath was sworn on the Sunday after the Easter holiday in Karlshafen in 1724. This was what the inventor made her swear,

> 'I, Anne Rosin Mauersbergerin, as I stand here, solemnly swear to you, my Master Johann Elias Orffyreus, by God the Almighty that I will not speak, write or show ill of you my past Master who stands before me, or uncover, divulge, speak or write anything of your deeds and omissions, art and secrets to any creature living or dead, but that I will keep each and every thing which I know or have seen or heard at your home, secret and hidden in myself from this hour onwards to my death and in eternity, as you request and demand of me. Indeed, I swear, once and for all, before God that I have not said or shown anything of your matters, machines and secrets to any person other than Mademoiselle . . . and that I regret with all my heart to have done so, particularly as my sayings were neither correct nor based on knowledge. So, please forgive me, it will not happen again or I shall be forever cursed, damned and lost before God and Man by all temporal and eternal judges if I should deliberately, knowingly and wantonly divulge, say or show anything of you and your secrets, arts and matters, or if I should try to bring shame and misfortune upon you, or injure your name, honour and life, or if I should speak, write or show ill of you or give anyone

> grounds to damage and destroy you; but that I will and promise to speak, write and testify well of you on every occasion and wherever I may be. And if I should break this solemn oath or if I should try to interpret it differently for malicious reasons or twist it or have it declared null and void, then God may plunge my soul into everlasting death without ever taking pity on me and deliver me into everlasting damnation as a soulless child, amen. I will be damned if I break this oath; I will be saved if I keep it and never break it. This has been sworn after due deliberation, sealed and affirmed with my mouth, hand and heart, of my own free will and without compulsion, in good will, by me, Anne Rosin Mauersbergerin, to you Johann Elias Orffyreus, my master. Amen, amen.'

This was a powerful and binding document — legally and psychologically. It ultimately had no effect. It is interesting to note that the inventor tried to cover all eventualities. He almost expects her to do all or some of the things he has tried to prevent her doing — and in fact she *did* do all of them. She spoke and wrote of his secrets; she attempted to dishonour his name; she testified against him. She tells us that she managed to get him to agree to pay her for the nightime turning by threatening to reveal all to the authorities, I wonder if she used another threat in 1724 to achieve some end of her own, and he responded by asking her to swear the oath. But she mentions that she has spoken of his 'matters, machines and secrets to Mademoiselle'. This may be a clue, because Madame de Langallerie was French and thus used such a title. Orffyreus discovers that his maid has been gossiping to Madame de Langallerie, and so he tries to undo the damage or prevent anymore by getting the maid to swear an oath of secrecy. He points out in the oath that what she has said is untrue; considering that the oath is between him and her and presumably was not intended to be seen by anyone else, one could assume that he was telling the truth. Why else would he say that she knew what she said was a lie, when he and she were the only one's privy to the remark . . . unless it was true, and she was lying? If she was telling the truth why force her to swear that what she said was a lie - thus negating the validity of the oath? Probably at this time he decided to have his whole family swear an oath of secrecy to enforce their silence. This is no different to any modern company, with a unique product, requiring its employees to sign a vow of secrecy prior to their employment. Strieder, introducing the above oath, takes on a doom-laden, ponderous tone thus, 'I would now like to reproduce the terrible oath which was sworn etc,.' Of course it was terrible — the inventor was trying to scare the living daylights out of the maid in the hope that she would hold her tongue.

Strieder uses the oath as evidence of Orffyreus' duplicity, but as we have seen, it had nothing to do with any fraud that he might have been involved in, and does not implicate him in any way. The swearing of oaths is and was a commonly used method of either getting a person to tell the truth — or to keep a secret. The inventor offers in his *Apologia* the chance for some people, who had requested it, to look into the hole in the side of his machine 'for a few ducats and the swearing of an oath' to keep silent about what they had seen. He even made Karl swear an oath before allowing him into the secret. It shows that Orffyreus placed a great deal of trust in the power of oaths — perhaps too much.

We can only speculate what led to Orffyreus' arrest, but it may well have stemmed from what ever it was that the maid said to Madame de Langallerie. Orffyreus had three enemies, Gärtner, Wagner and Borlach. Gärtner died on the 2nd February, 1727, so may not have had much to do with the arrest, although he may have been initially involved in the planning of it. He was seventy-three years of age and still harbouring a grudge against Orffyreus. Wagner certainly showed plenty of spleen against Orffyreus and he may well have been acting behind the scenes. Then there was Borlach, he of the infamous cartoon. In a pamphlet he issued in 1716 he stated that Orffyreus' perpetual motion machine was impossible, and there could never be such a thing. He even calculated how long it would take to carry the whole world several feet higher using the Orffyrean machine — and even if the Orffyrean machine was not moved by a cord, he opined, it was absurd anyway.

The actual incident that led to Orffyreus' arrest may have been caused by the maid's departure from his employ. No explanation is given as to why she left, but it may have had something to do with her mistress's death in 1726. Perhaps she wished to return to her original home of Drebach near Annaberg, where her family may still have lived. She was thirty-eight years old, but her marital status is unknown, and it is possible that she could have obtained other employment at court. She might even have dreamed of a future with Orffyreus himself, but that is pure speculation. There has to be some explanation for the obvious spite with which she ruined her previous employer's reputation. Orffyreus married again, but who and when is a mystery, perhaps this event turned the maid from a gossip to a vindictive fury. If he did re-marry shortly after his wife's death, the maid could have felt angry not only for her own exclusion, but alternatively for her late mistress's early replacement.

Strieder reports that,

> 'On the occasion of her departure from his employment, Orffyreus took her to one of the most remote rooms in the house, firmly locked

> the door, placed two loaded pistols on the table and ordered her to swear another terrible oath, adding that if she refused to do so, he would shoot her through the head there and then; and when she replied that that would result in his death as well, he said that that would not matter since he would have to die anyway. She then suggested that a new oath was not necessary since she had already sworn to stay silent, to which he replied that, if she refused it would seem as though she wished to ruin him and tell everybody all that she knew and asked her not to make such a fuss since . . . there was no secret; he also tried to impress on her that turning the wheel had been done only to prevent the machine suffering internal wear and tear, and warned her that she too, would be punished since she had assisted in the work. Albeit reluctantly, but to save her life, she agreed to his demand and swore the oath, although she did not feel bound by it as it had been extracted from her by force. Prior to her leaving, he gave her a long speech, telling her to keep quiet and reminding her of her oath'.

Again one can sense questions and answers. Why did she agree to the swearing of a second oath? To save her life, there were two loaded pistols on the table. Why did she not keep the oath? Because it was forced from her. Did he ever admit to her that he was involved in a fraud? Did she not ask why she was being made to turn the wheel? Does she realise that she is implicated in the fraud? He told her that turning the wheel was merely saving wear and tear. One can also sense that she was no push-over. When allegedly presented with the sight of two loaded pistols, the maid does not dissolve in floods of tears, in fear for her life; she calmly retorts that if he kills her he will face death too. Further on, their roles seem to change, and he is the one doing the pleading. I am sure that the maid's original claim that the wheel was turned by a secret mechanism hidden in the posts was completely fictitious in which case, as we only have her word for it that any of the above happened, I am inclined to think that most of it is false, although there may have been elements of truth in it. For instance, I can well believe that Orffyreus might have got her to swear a second time, and that suggests that although he believed in the efficacy of oath taking, he did not trust its enduring potency as far as the maid was concerned.

So — if the maid was lying about the deception, what was it that the inventor was afraid she would reveal? If he was a fraud — she could reveal that. But if he wasn't — it might have been the actual secret of the machine that he was trying to protect. Perhaps he was afraid that she had accidentally seen the interior of the machine, and even if she did not understand it, she still might have been able to give a good enough description of it to allow someone else to work out how it was done. Having worked for the inventor for around fifteen years, it would be surprising if, in all that time, the maid had not stumbled upon a half-finished machine, or one that had had the cladding removed, or one that was in the process of being 'smashed', or seen delivery of some special part required for the wheel — and then drawn her own conclusions about how the machine worked.

Returning to Strieder's account we learn that ' . . . although the Orffyrean misdemeanours have not remained hidden, they were never prosecuted either.' Evidence was taken but the prosecutors were unimpressed and Orffyreus was allowed to go free. The evidence for the prosecution appears impressive but is easily dissected and rejected. Unfortunately for Orffyreus, the arrest, examination and release without charge was all that his enemies required. Once rumours of his arrest for fraud had got round, no one ever believed in him again. It became impossible for him to promote his machine seriously again.

What of Karl while all this was going on? Surely he would have something to say in defence of Orffyreus? He, after all knew the truth. In 1727 in a letter, he wrote,

> '. . . We do not give the supplicant any credence... because We are otherwise persuaded.'

A dignified and emphatic, if brief, response. It is as though the matter were unworthy of serious consideration, and should be treated with the contempt it deserved. Unfortunately for Orffyreus it had to be treated with considerable alarm, as the effects of the maid's claims crept ever outwards like the ripples in a pool.

Although a super-wheel was never constructed there is evidence that a smaller one existed in 1727. In that year, in one of his annual panegyrics to Karl, the inventor complains about the outcry against him and his invention, and adds that Maximillian, one of Karl's sons, has been to see him and his wheel in Karlshafen. So years after 'sGravesande's inspection of the wheel at Kassel in 1721, there existed one at Karlshafen. And there is additional evidence to back this up.

One of professor 'sGravesande's correspondents was a Swiss called Jean-Pierre de Crousaz (1663-1750). He was known as a theologian, philosopher and controversialist, whose greatest importance lies in his letters to a wide range of correspondents. He was a professor at Lausanne from 1700 to 1724, and returned there from 1738 to 1749, but during the intervening years, as well as holding a chair at Groningen in the Netherlands, he was tutor to Karl's grandson Frederick. He wrote numerous mathematical and philosophical works and attempted to refute the doctrine of Leibniz. Arriving at Kassel in 1726, he must have become instantly aware of the rumpus caused by Orffyreus' maid. Perhaps he was informed - inaccurately, as we shall see — by the Marquise de Langallerie. Crousaz wrote to Professor 'sGravesande from Kassel, on the 3rd February, 1729, and summarised his opinion thus,

> ' . . . Firstly, Orffyreus is mad.
>
> Secondly, it is impossible that a madman can have discovered that which such a number of clever persons have searched for without success.
>
> Thirdly, I do not believe in impossibilities.
>
> Fourthly, we can easily imagine that persons keep a secret, from which they are to receive benefit; but this fellow, hoping only to gain reputation, allows this to be tarnished by an accusation which he has it in his power to disprove, if false.
>
> Fifthly, the servant who ran away from his house, for fear of being strangled, has in her possession, in writing, the terrible oath that Orffyreus made her swear.
>
> Sixthly, he only had to ask, in order to have this girl imprisoned, until he had time to finish his machine.
>
> Seventhly, they publish that the machine is going to be exhibited, when suddenly those who advertise it become silent.

Eighthly, it is true that there is a machine at his house, to which they give the name perpetual motion; but that cannot be transported, it is much smaller, and it differs from the first, in that it only turns one way.' . . '

Professor Allemande reproduced 'sGravesande's reply to this letter. It went as follows,

> 'I have deferred writing to you, until I had found the paper which I wrote the day after I had examined Orffyreus' machine; for although I remember well all that passed, I believe that a paper, written the day after the examination, and communicated to His Serene Highness the Landgrave and to all those who were with him, must have more weight. It was for this reason that I wished to know how I had expressed my opinion.
>
> 'This is what I have heard; they say that a servant under oath, turned Orffyreus' machine, she being in an adjoining room. I know perfectly well that Orffyreus is mad, but I have no reason to think him an impostor. I have never decided whether his machine is an

imposture or not, but this I know, as certainly as anything in the world, that if the servant says the above, she tells a great falsehood.

'His Serene Highness the Landgrave, in the presence of the Baron Fischer, Architect to the Emperor, and other persons, ordered at my request, that the machine be dismounted from its supports, and we saw the bearings uncovered. I examined the bearings on which the journals rested and in that examination there did not appear to be the slightest trace of communication with the adjoining room. I remember very distinctly the whole of the circumstances of that examination which put Orffyreus in such a rage with me that, the same day, he broke his machine into pieces, and wrote on the wall that it was the impertinent curiosity of Professor 'sGravesande which was the cause. I could read this inscription myself one year later, when I was in Kassel'.

'The result of my examination is clearly revealed in the letter that has been published in *Mercure Historique*, September issue of 1721. I was told of many circumstances surrounding the sworn testimony of the servant, but I do not see any difficulty here. As for the machine, I do not rely on the word of a servant. Perhaps when she turned the roasting spit for her master she believed that she was turning the perpetual motion machine.'

'If you hear of any further specific details of this affair, it would give me great pleasure if you would describe them to me'.

After giving the above letter, Allemande makes a few comments himself.

It is difficult to decide what to think about this machine. It appears to me that after thoroughly examining the pros and cons of Orffyreus' machine, we must conclude that,

1) Orffyreus was evidently mad, as Ms 'sGravesande and Ms. de Crousaz both affirm; his breaking his machinery at different times with neither reason nor necessity proves this. But his was a sort of madness we don't often see; a mania fixed only on certain objects, which merits the name of fantasy or whimsy — this kind of mania is often accompanied by much ingenuity, and when persons of this disposition apply themselves solely to one subject, as it appears he did, it is not surprising to find them making discoveries which have escaped the sagacity of wiser people.

'Thus I do not agree with Ms. de Crousaz, that it is incredible that a madman such as Orffyreus, should have found out something that learned men have sought unsuccessfully. Furthermore, he is wrong to say that Orffyreus could not expect anything but reputation from his secret; he expected a considerable profit, because he demanded 100,000 thalers for the secret.

2) No exterior agent moved the machine; if it were a servant that moved it, would not this have been apparent to eyes so searching as those that made the examination, or to the Landgrave, who had seen the inside of the machine? Besides, how can one believe that a wheel of so great a volume could have been moved by such a cause, a cause which would only act on the bearings crossing the supports, and which must have been so small as to have escaped the most rigorous examination?

3) Supposing the servant has not been bribed to make a deposition against Orffyreus, what does her testimony prove? Only that her master made her believe that, by turning a little wheel, she moved the whole machine — and we can imagine a singular character such as he was might have done this to baffle the curiosity of those who sought to penetrate his secret. This I heard Ms. 'sGravesande say many times.

4) We must admit that the machine was a very remarkable mechanical phenomena. This should be kept in mind when we express an opinion about it. It would be as rash to assert that it was a perpetual motion as to regard it as just an imposture moved by some external agent'.

Going back, first of all to de Crousaz's list, summarising his opinion of the affair, I was struck by the puerile quality of his much vaunted powers of criticism. He was widely acclaimed as a sceptic, but if this was an example of the kind of critique for which he achieved fame, then his fame must have rested on the volume of his polemic rather than the content. One biographer states that he was 'chiefly distinguished for his treatises on Logic', and one can only assume that on this occasion he was suffering from an off day — or he had some other ominous reason, not unconnected with Orffyreus' enemies.

To take the first point, he says Orffyreus is mad. There is no elaboration, and it takes the efforts of both 'sGravesande and Allemande to extract something worthwhile from this statement. Orffyreus was not mad, not certifiable; he was obsessed and eccentric, that is all. As Allemande pointed out, it was this single-minded approach that gave him the advantage over fellow perpetual motionists. It is worth pointing out though, that many perpetual motionists both now and in the past display an equal obsession in their search for the secret, but fail to find it. Point two of de Crousaz's list claims therefore that it is impossible for a madman to have discovered what many clever people had failed to discover. That has already been covered, and was really a part of his first point.

Point three is not even a point. He, de Crousaz, does not believe in impossibilities. Does anyone? Allemande and 'sGravesande ignore this point. It is beyond comment.

Point four, was adequately dealt with by Allemande who reminded us that Orffyreus could expect a lot more than a boost to his reputation, i.e. 100,000 thalers. It seems very odd that de Crousaz should have been unaware of the cash figure that the inventor asked for the secret of his machine. De Crousaz says that the inventor can easily prove that his machine is not a fake by revealing the secret; but that was probably the point of the whole exercise, and Orffyreus must have considered the possibility, and made sure that he did not give anything away; perhaps trusting that the Landgrave would come to his rescue.

Point five offers us the information that the servant ran away for fear of being strangled, and that she has the oath in her possession that Orffyreus made her swear. I have been unable to find any reference to the maid being fearful of being strangled, although it is true that she may have feared being shot by Orffyreus, as he is alleged to have had a pair of loaded pistols on the table at the time of the swearing of the oath. It would not be surprising if anyone in his position threatened to strangle the maid in the particular circumstances that pertained, but that is not de Crousaz's main point. The fact that the maid has this terrible oath in her possession is proof enough for de Crousaz that Orffyreus is guilty. It is noticeable that every time the word oath is used, it is preceded by the emotive 'terrible'. The reasons for Orffyreus' use of the oath has been discussed, but I do not see that one can assume, as de Crousaz has done, that the maid ran away for fear of being strangled or shot. She was already leaving for what ever reason, and that is why the oath was required, and if loaded pistols were in evidence, then that was just to lend gravity to the occasion, and to instil a little fear into the maid.

Point six suggests that Orffyreus only had to ask to have the girl imprisoned. That is unlikely. For a start we know Karl's mental and physical condition was deteriorating. His mistresses were pressurising him, his sons were carrying much of the work load, and if those who brought the accusations to the public's notice were so powerful that they could secure his arrest, it is unlikely that those same people would have acquiesced to a request from Orffyreus for the maid's arrest. The courts of those days were hotbeds of intrigue. Powerful factions strove for supremacy in obtaining the Ruler's ear. Time after time, throughout history, the Ruler's favourite has been shown to hold the keys to power. The Marquise de Langallerie shared that position, and if she wanted Orffyreus' head on a plate, metaphorically speaking, her status allowed her to demand it. Included in this point is the statement that imprisoning the girl would allow time for the wheel to be finished. The implication being that Orffyreus had at least started on a new wheel, perhaps the planned largest ever. Was it the timely action by the maid that allowed the arrest and at the same time prevented the inventor from finishing the wheel?

Point seven informs us that a new wheel is to be exhibited, then they are suddenly silent. No wonder — the inventor has been arrested. How can they continue with the exhibition of the machine, if the inventor is in prison?

Point eight is the most extraordinary one of all. De Crousaz admits that there is a machine at Orffyreus' house, which is called a perpetual motion machine, but it cannot be transported and can only turn one way. Does it matter if it cannot be transported? Does it matter if it can only turn one way? The fact that the first wheel, at Gera, could turn at all, of its own volition, aroused great admiration. This later machine was at least as good as the original, and yet de Crousaz is almost sneering in his contempt for it. He ignores the fact that the machine complies with the requirements of a perpetual motion machine; that it should turn under its own internal power, continuously.

Why did de Crousaz attack 'sGravesande's evidence? The tone of his criticism must have been offensive to the professor, and it would have been easy for him to have either ignored the implied criticism of his own examination, or to have meekly accepted that he could have been mistaken. He does neither, because he is satisfied with his findings; and of course the points raised by de Crousaz are simple to refute. Being tutor to Karl's grandson must have brought de Crousaz into daily contact with the Marquise de Langallerie, and she could well have given him a biased account of the Orffyrean affair. She, possessed of great beauty and charm, flattered him with her attention and persuaded him to adopt her viewpoint. De Crousaz had spent much time debunking Leibniz' theories, and it may have been logical for him to take the opposing view of Leibniz who supported Orffyreus, even though he believed the inventor was mistaken in describing his machine as a perpetual motion. Professor 'sGravesande stuck to his guns and declared in no uncertain terms that the maid was liar. There is no record of a reply from de Crousaz.

The possibility that his secret might be uncovered had obviously occurred to Orffyreus. He seems to have kept his machines permanently disassembled except when being exhibited. Thus even though he was arrested, no complete machine was found and his wheel's hidden mechanism remained hidden. Further precautions were taken later because on the front cover of his *Maschinen Tractate*, are written the following words in his own handwriting,

'Further demonstrations regarding the possibility and impossibility of perpetual motion.
N.B. 1st May, 1733. Due to the arrest, I burned and buried all papers that prove the possibility. However, I have left all demonstrations and experiments since it would be difficult for anybody to see or learn anything about a perpetual motion from them or to decide whether there was any truth in them because no illustration by itself contains a description of the motion; however, taking various illustrations together and combining them with a discerning mind, it will indeed be possible to look for a movement and, finally to find one in them'.

After Karl's death Orffyreus continued to publish pamphlets containing new ideas. In 1738 he announced three new devices: 1) a fountain which leapt continuously from still waters; 2) a large musical organ which played by itself, and 3) his 'Orffyrean ship'. or 'special preservation machine' which was designed to save life and goods following a shipwreck. It is not difficult to see that all three machines would be unable to function without the presence of his perpetual motion machine. The inventor had not given up on his machine, but rather, had found an alternative way of marketing it since the words 'perpetual motion' might rouse memories of his arrest. The fact that he continued to try to sell his machine under a different guise, lends credence to my assertion that his machine was a real invention and not just a subterfuge to make money. Nothing, surely, would have persuaded him to once again attempt to interest the public in schemes that were clearly related to his original enterprise, unless he really had discovered the illusive secret of perpetual motion.

Strieder attempts to score more points off Orffyreus by drawing attention to the fact that the inventor had claimed in his 1738 pamphlet that his submarine was perfect and incapable of improvement, when a later version, in 1739, presented improvements to the original. In fact all machinery is made to the best possible specification at the time; but is always capable of improvement. Orffyreus was merely using a modern marketing ploy to sell his machine. Modern day equipment is always described as 'the latest model' or 'at the forefront of technology' and yet improved versions always follow as the night follows day.

Little else is known of Orffyreus' life until two years before his death. I have copies of many letters written to him and by him in the intervening periods, but they are mostly written copies, kept for his own benefit, and barely legible, unfortunately. It is evident from them that he continued to correspond with a great many people throughout Europe, but the details remain hidden. In 1743 he travelled to the Brunswick region. Karl I, Duke of Brunswick-Wolfenbüttel had succeeded to the Dukedom in 1735, aged only twenty-two. Two years previously he had married the daughter of Frederick William I of Prussia, and sister of the future Frederick the Great of Prussia. The young Duke, Karl I, was modern in thought and planned to industrialize his country. He imported experts and one of these was Orffyreus, who had apparently lost none of his powers of persuasion. Initially he had been involved in setting up a marble slab factory and had also advised on the establishment of a Russian and Moroccan leather factory. The notes and diagrams of this venture are still in existence in the inventor's own hand.

His final enterprise for which his advice was sought was the design and erection of a windmill. Being Orffyreus, a standard windmill would not do. His idea was to construct a horizontally sailed windmill, one with a vertical axis, that would work irrespective of the wind direction. Such a thing had been known for centuries in Persia, but was unheard of Northern Europe. The site for all these enterprises was the small town of Fürstenberg, and Orffyreus planned to place the windmill, which was to be used for the production of flour and oil, on the cliffs of the nearby Kathagen Mountains in order to take advantage of the up-winds off the river Weser. However Karl I, assigned him a place within the town itself, too sheltered to allow the new windmill to operate effectively.

The inventor began to build a massive two storey sand-stone construction. On top of this he placed a further two and a half floors, half-timbered. The walls were huge in order to support the weight of the sails, and to provide a firm anchorage for them in any kind of wind. The doors were edged in red sandstone, and the walls were plastered white. The main door had the monogram of the Duke, two entwined Cs, on the lintel. This is as far as he got. On the 30th November, 1745, at the age of sixty-five, the inventor of perpetual motion fell to his death.

His fall has been attributed to a variety of different causes. Some say that the unsuitability of the windmill site preyed on his mind. Was he building a monster windmill that would remain motionless? Did he jump to his death? Others blame the extreme poverty in which he lived at this time. He had remarried and had young children and a wife he could not support. Lack of food and ill-health coupled with hard work and the inclement weather of late November may have contributed to physical weakness, leading to his fall.

There exists a sadly evocative letter which the inventor wrote at the end of April, 1745, only a few months before his death, from his home in Karlshafen, in which he pleads, wretchedly, for some small help for him and his family. But as well there is, intriguingly, a reference to one more perpetual motion machine which he has been ordered to finish — and also to a steam engine. The letter is addressed to Baron Schrader von Schliestedt, Karl I's right hand man.

'Sir, Baron and Highly Respected Court Councillor,

'I have written often and many times to you, Highly Honourable Sire, but I have not received any answer. Meanwhile, I lack strength and will, through hunger, grief, frost and chill, because for a long time I have not received a single piece of wood from the District Magistrate. Indeed, I have often had only dry bread to eat and water to drink. However, at last, thank God, I have had the good luck to finish preparing the model of perpetual motion which was commissioned from England, as demanded by the Senior District Magistrate, von Mannsberg. It was ready at midnight on April 14th, and on the same day I had travelled to Karlshafen, borrowing two gulden from the Rev. Pastor. On the 15th of the same month I had to return home from church to go to bed, because I suffered a violent attack of colica bilosia flatuente. It was the result of my cold, miserable life, and general anxiety about the Fürstenberg project. It afflicted me for 8-9 days, so that I was sure that death was near. All this was also accompanied by the effects of an accident suffered during the work at Fürstenberg. However, my God helped me to bring such a sudden change using only the good medicine of my mind, that I hope to return to Fürstenberg this week. To Him only be thanks and praise.

'... it would be of great relief to me and help my health if I were to receive a pleasing reply from you, to my letters. Moreover, it would also please God, because nothing else has produced so much willingness in my heart, to be of service and to have honest intentions, and so there is much that I deserve. I would wish that Your High and Honourable self were to be made aware of and have insight into how matters really stand. You have the power to affect my life to an appreciable degree, either prolonging it or shortening it. Because if, in Fürstenberg, things continue as they have done, then I am absolutely certain to die of various heartbreaking anxieties. This could happen at any time and I am ready and willing; because I cannot live here any longer in absolutely unbelievable grief and privation, with mounting debts, and no money to repay the debts already incurred; and, in addition to the pain, trouble and other difficulties experienced with the mechanical construction, which was started so late, and which had had to be begun anew, there is nothing to eat for me and my family. The District Magistrate was unwilling to lend me one barrel or even sixteen pecks of grain, and so things have been impossible for a long time. It hurts my heart when I see and hear, sometimes daily, what is done or said against my Christian honesty, and there is nothing I can do, but only to bear it patiently. It all effects my spirit when rude comments are made in front of strangers, and am I to believe that you are my avowed foe? And this because steam engines have not been made! And yet when I ask where is the solemnly promised wood and other things for this, then they keep silent, but behind my back they repeat the same. Recently I asked the District Magistrate for just five thalers for expenses and I have received nothing. But he says, on the other hand, he had offered me ten, twenty thalers, and so on. I should not have to suffer like this at my time of life. There is so much more that I could say, but why bother? In short, I will state my honesty once and for all, and as well, recommend myself as a poor and weak man, who seeks your Honourable protection with due respect and honest modesty. Your most humble servant.

Karlshafen, April 26th, 1745.'

Orffyreus' problems were insurmountable. Worry over the windmill; his lack of success with a steam engine; illness and mounting debt — these factors combined with an accident at work and the malicious insinuations about him, spoken loudly within his presence seem to have led to his death. I think it very likely that his death was not altogether involuntary. He may not have jumped, but I suspect he did not care if he lived or died, and with winter approaching, conditions for himself and his family must have been intolerable. Perhaps he was working on the top of the windmill, and a particularly strong gust of icy wind threatened to dislodge him, and on this last occasion, it was easier to just let go.

It is a sad letter and the treatment meted out to him seems unfair. He ends his letter by restating his honesty, and if his invention was genuine — doubly sad. It is worth remembering that he steadfastly maintained that he was innocent of any fraud throughout his life.

The reference to the steam engine implies that one was required by Karl I, and as he was married to the sister of Frederick the Great, King of Prussia, he probably got the idea from him. Frederick wished to buy steam engines for the mines in his country, and the original drawings made for him, still exist.

The model of the perpetual motion machine which was ordered to be got ready for the commission from England is an intriguing. Von Mannsberg held the title to the land at Karlshafen, but worked in England as assistant to von Hatorff, and later to Ernst von Steinberg, both men being successively responsible for overseeing George II's German possessions. The new action to secure an Orffyrean wheel may have been initiated by von Mannsberg reading Desagulier's newly published lectures on perpetual motion. The curator of experiments at the Royal Society, of which von Mannsberg was an FRS, published his course of lectures, given years before, in 1745 and the reference to John Rowley and the Orffyrean wheel, may have struck a chord with Orffyreus' landlord. The name 'Orffyreus' would certainly have jumped out of the page at von Mannsberg, and he might well have recollected that the inventor working at Karlshafen had the same name. The thought that the man of whom Rowley had spoken of so convincingly was living and working on his own land back at home, must have roused his curiosity. He wrote directly to the inventor and commissioned him to build a new wheel for him. Whether he would have purchased the machine is debatable, in any case events overtook his plans. Fresh scares about another Jacobite rebellion surfaced in this year, and advice was given against travelling in Europe until the threat had passed. By then, of course Orffyreus was dead.

The misery in which Orffyreus finds himself is almost palpable. It is impossible not to feel sympathetic to his plight, especially considering how high his status had been, and how low it had sunk. There is little doubt that the windmill would have been a success if it had been built where he had intended. But it was not wasted. Two years after his death, the Duke's master of hunt, von Langen, who worked in Fürstenberg, offered his master some plans which he had drawn up which proposed the setting up of a porcelain factory. Von Langen had the hunting lodge in mind for the factory, but although the Duke gave his blessing to the venture, it was not the hunting lodge that he allotted to von Langen on the 11th January, 1747, but the still unfinished windmill. It was from this small beginning that the famous Fürstenberg porcelain industry sprang. The original building that was constructed by Orffyreus has been incorporated into a folklore museum for the tourist industry.

According to Strieder, Orffyreus was buried as he wished, in the family grave in his garden at Karlshafen. However local rumours suggest that he was in fact, buried in the forest nearby, which was commonly done at the time. Certainly, to date, there is no evidence of his grave, despite the fact that a possible candidate for the original house has been found. The finding of this gravestone could have considerable importance for reasons which I shall go into shortly.

The widow of the inventor, his second wife, had no savings or income, so she received a weekly handout from the governmental administration of one thaler, twelve groschen and some firewood. It is said that she offered the Duke the inheritance of her husband, which consisted of a model of the perpetual motion machine, which had predictably, been taken apart before his death. There was a collection of engravings, some mechanical and optical instruments and many books which apparently were of no scientific value. The books may have been Orffyreus' reference books for the cure of various ailments, plus of course his *Maschinen Tractate,* which he had once planned to publish.

Thus ended the life of a man who had wanted to achieve so much. History has accorded him a place among the frauds, fakers and fabulists that enliven the records gathering dust in various museums world wide. Was he a swindler? Did he fool everyone with a machine which was just a clever device concealing a secret connection to an adjoining room? A room in which sat a paid operator turning a handle which rotated the huge wheel. Could this 'madman' have carried on his deception for thirty-three years, until the day he died, and even after his death defy those who accused him of a cunning deception? Is perpetual motion possible? If it isn't, what *had* Orffyreus discovered? Why has it not been re-discovered? How could such a machine help the inhabitants of our planet in the twenty-first century? These questions need answering, but before we address them, there is one more puzzle to solve, and it is a conundrum that could lead to a solution to the above questions. Why did Johann Ernst Elias Bessler change his name to Orffyreus?

Chapter Eleven

WHY ORFFYREUS?

Johann Bessler has always been known as Orffyreus, and although the method he used to arrive at the name is well established, no one, it seems, has ever speculated that there might have been a serious reason for adopting this alias. In investigating the reasons for using a different name to the one he was born with, a certain amount of speculation is unavoidable, but when added to the known facts, it becomes increasingly credible that Bessler did not just change his name on a whim — he had an excellent reason for doing so.

In his *Apologia*, Orffyreus tells us that Wagner wanted to know why he called himself by a name other than his own. Wagner implies that it was for the purposes of criminal deception — a means by which the inventor could escape incognito from various local authorities who were after him for some illegal activity or other. This accusation does not stand up to examination. If the inventor wished to remain anonymous or incognito, a name like Orffyreus was about the least likely one to choose. He would have used a common name in order to remain hidden. Others have suggested that he wanted a stage name, so that people would remember him and his wonderful cures for various ailments. This latter suggestion may be true but it is not the whole story.

It was a common practice at this time to Latinise the endings of personal names to give them a spurious classic appeal, hence Wolff became Wolffius, Mencke became Menckius, and Papin became Papinus. The name Orffyreus was originally Orffyre, but was Latinised for fashion's sake, and to add a touch of sophistication; but to get to the beginning of this mystery we need to consult our old friend Strieder again.

His very first words on Orffyreus tell us what we want to know, he says,

'The true name of this highly infamous man was really Bessler. His Orffyre was formed by superimposing the letters of the alphabet in the following manner:

a	b	c	d	e	f	g	h	i	j	k	l	m
n	o	p	q	r	s	t	u	v	w	x	y	z

'By substituting the 'b' for Bessler, in the top line for the 'o' for Orffyreus in the lower line, and the 'e' in the top line for the r in the lower line, and so on to the end, sometimes having to interchange the top with the lower line, the name Bessler is changed to Orffyre. The result was then Latinised. But why? We have established that it was unlikely that it was to provide an alias for activities on the edge of the law. What other reasons could there be for such a change?

One account of the legend of Orffyreus' wheel, suggests that the letters of the alphabet were arranged in a circle, and those diametrically opposite to the ones spelling Bessler were picked out and spelled Orffyre. This is an alternative to the above system, and it does work. The same author goes on to ask if this pattern in some way symbolised the wheel and how it worked. If one accepts this as a possibility, it is a small step to wonder if any other hidden symbolism exists within the author's other works — perhaps even the secret he possessed. We will return to this thought shortly. Others have implied that Orffyreus was steeped in cabalistic doctrines and that he was using an ancient cabalist's formula to arrive at a name that would have meaning to those in the know. How he arrived at the name is interesting but not as important as finding out why. A clue to this may lie in the quotations used on the front page of the *Apologia* of 1716 and his later publication *Das Triumphans Perpetuum Mobile Orffyreanum* of 1719. In the first case, at the bottom of the title page he includes a quotation from the New Testament, Matthew XV, vs 16,

> 'Are ye also yet without understanding?'

The full quotation as it appears in the Bible, is part of a conversation between Jesus and the disciples, and Orffyreus repeats the quotation on the very last page of the same work, adding the following comment,

> 'Were I to reveal the details of my art, the Devil might enslave you'.

This does not seem to be a particularly relevant quote to use. He might have used one that urged belief in God or work or honesty; or he could have used one that conveyed the everlasting nature of his machine. A hundred more suitable passages were available to Orffyreus, in the Bible - why did he choose this one? Was he implying that there was actually something significant to understand? His comment at the end of the repeated quotation actually answers the question — if he revealed his secret the Devil might enslave you. His next publication also contained a quote from the Bible, and seemed to enlarge on the first one. This one, from the Apocrypha, Tobias 12, vs. 8,

> 'The secrets of Kings and Councillors man should be silent on, but God's works he should openly praise'.

This time there is actually a secret mentioned, but it is one known only to Kings and Councillors. Orffyreus is a Councillor of Commerce, and Karl is a ruler, are we supposed to make these associations? Certainly they were the only two people who knew the secret of Orffyreus' perpetual motion. The first quotation has an additional curious facet to it. 'And are ye yet without understanding' is written in eighteenth century German, as is the whole work, but there is a unique difference to this sentence. Unique to Orffyreus, that is, for he uses it more than once. The type is in a heavy gothic script, and it actually appears like this,

'SenDIheDan a VChnoChVnVerstanDIg', (see photo). I did not notice this oddity for a long time, and when I did, I thought it was just a curiosity of the printer. But having conceived the idea of a possible hidden message when pondering on the reasons for Orffyreus' name-change, I extracted the capitals and immediately realised that they were Roman numerals.

D I D V C C V V D I

D=500, I=1, V=5 and C=100. When added up the figures amounted to 1717, the year of publication of the pamphlet, and the construction of the inventor's largest wheel. The result was at first disappointing. Was Orffyreus playing games with numbers and letters? I was afraid so to begin with but then when I reflected on the point I began to understand. There are hundreds of Latin words in the *Apologia*, but nowhere else in that particular publication does he use this little trick, although he does use it again in his next tract. This particular perplexity would have been more obvious to readers of those days who would have been familiar with the gothic script currently in use then, so it could not reveal anything of significance, but I believe it was another tiny clue that there was more to be seen than met the eye in Orffyreus' published works.

If the inventor wanted to lead an investigator to his secret after his death,

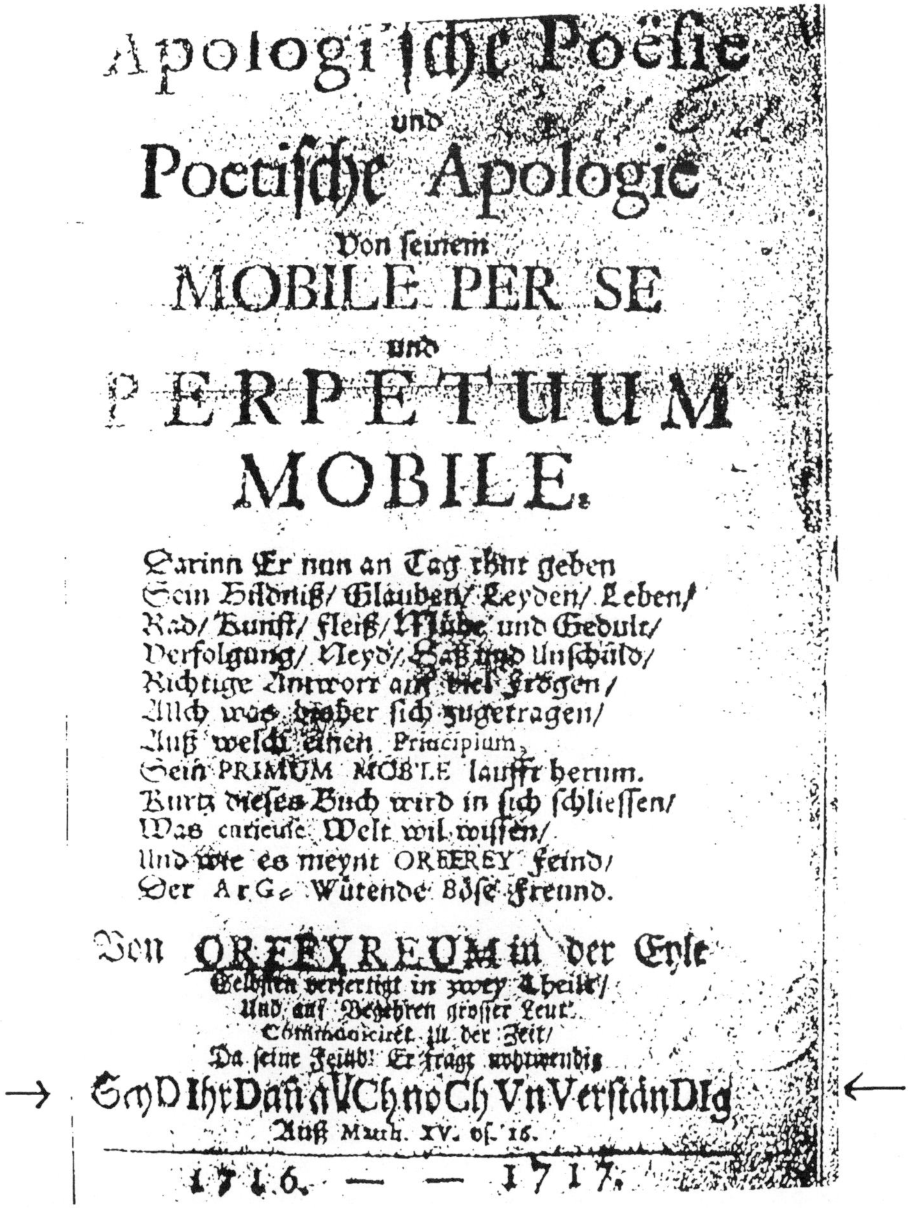
Apologi'sche Poësie
und
Poetische Apologie
Von seinem
MOBILE PER SE
und
PERPETUUM
MOBILE.
Darinn Er nun an Tag thut geben
Sein Bildniß/ Glauben/ Leyden/ Leben/
Rad/ Kunst/ Fleiß/ Mühe und Gedult/
Verfolgung/ Neyd/ Haß und Unschuld/
Richtige Antwort auf viel Fragen/
Auch was bißher sich zugetragen/
Auß welch einen Principium,
Sein PRIMUM MOBILE laufft herum.
Kurtz dieses Buch wird in sich schliessen/
Was curieuse Welt wil wissen/
Und wie es meynt ORFFEREY feind/
Der ArGe Wütende Böse Freund.
Von ORFFYREUM in der Eyle
Selbsten verfertigt in zwey Theile/
Und auf Begehren grosser Leut
Communiciret zu der Zeit/
Da seine Feind' Er fragt nohtwendig
→ SeyD Ihr Dann aVCh noCh VnVerstanDIg ←
Auß Matth. XV. v. 16.
1716. — — 1717.

The above detail from the first page of Orffyreus' 'Apologia' shows the 'Latin Capitilising' of the Gothic script, which when added together vertically produces the total, 1717.

then there are only a limited number of places where he could place the clues. The most obvious place to start looking would be his published works. They could not be too obvious or someone might discover the secret too soon, while he still lived. A number of these small hints could go unnoticed, and would only acquire significance if the inventor had, after he had died, left some pointer to the fact that there was a message waiting to be found. What exactly would the message reveal? Only the secret of his wheel's construction could be worth hiding.

If various small clues seem to imply that Orffyreus had left a message to posterity, how would he have attracted attention to this in the first place? The clues were, after all, designed to be overlooked while he was alive, so it would have to be something that only came into public view following his death. It is clearly indicated in several places that Orffyreus sought and obtained permission from Karl the Landgrave, to erect a family grave for his family and himself, in the garden of his house in Karlshafen. The house given to him by Karl. From his *Apologia* it is known that Orffyreus had the requisite skill to design, construct and erect his own tomb or vault or, perhaps just a tombstone. Strieder says said that when he died, his body was returned to Karlshafen and buried in the family grave there. Whether it was or wasn't, the inscription which I am sure Orffyreus had carved in the fabric of the memorial might have included a message to the reader who had more than a casual interest in his life. I believe that the inventor had designed and inscribed his own tombstone with the purpose of leading those who came to see it, on a journey of discovery ending with the revelation of his secret — the means of constructing an Orffyrean perpetual motion machine. Unfortunately the knowledge of the whereabouts of Orffyreus' grave has been lost. However, having accepted the possibility of there being some kind of coded instructions hidden within the Orffyrean texts, what method could the author have used to secrete his message?

The history of codes is an ancient one, and it is not necessary to go back to the first known codes, however there is an interesting aspect to one of the earliest examples of codes that has a direct bearing on the problem we are attempting to solve. The Bible is one of the places where one would least expect to find evidence of the use of codes, but it is well-established that they were used. The words which were enciphered were personal names and the enciphering could have been done to hide the real names of the people being described.

The original language of the Bible was Hebrew, and some names were transposed into different names. The transformation resulted from the application of a traditional substitution of letters called 'atbash', in which the last letter of the Hebrew alphabet replaced the first, and vice versa; the next to last replaced the second, and again vice versa and so on. This would be the equivalent of a=z, b=y, c-x until z=a was reached.

a b c d e f g h i j k l m n o p q r s t u v w x y z

z y x w v u t s r q p o n m l k j i h g f e d c b a

Another system also used in the Old Testament was known as the 'albam' method. It split the alphabet in half and equates the two halves.

a	b	c	d	e	f	g	h	i	j	k	l	m
n	o	p	q	r	s	t	u	v	w	x	y	z

Thus the first letter of the first half, in English script the 'a', is substituted for the 'n' in the second half, and all the letters in one line are substituted for those in the second half and vice versa. That is of course the system used by Orffyreus. Hebrew literature also records a third method called 'atbah', which is similar to the above systems, but is based on Hebrew numbers. Hebrew numbers, like Roman numbers, used letters from their alphabet. The numbers one to nine were substituted with other numbers from one to nine, ensuring that the substitute plus the number being substituted added up to ten. Thus the number one would be substituted for the number nine, two by eight and so on. Larger numbers were treated the same but paired with numbers totalling one hundred.

Orffyreus undoubtedly used the albam method to construct his name, and the idea must have come from the old Rabbi whom he visited in Prague. Orffyreus says in his *Apologia*, that the Jew taught him 'Hebrew, hieroglyphics, the language of nature and the writings of Angels'. The Rabbi was certainly familiar with the various Hebrew cyphers and taught them to his helper.

At first sight this does not help us, other than explaining how Orffyreus arrived at his name. Other codes must have been used which could be innocently included in the text without unwanted attention being drawn to them. If this is the case, the name 'Orffyreus' plus the meaningful quotations appearing on the title pages of his books is intended to be a pointer to further hidden information for future readers There is another eccentric factor which appears in *Apologia* which undoubtedly holds an encoded message

Apologia is produced in two parts, and at the end of the first part the author issues his 'Declaration of Faith' as he calls it. This remarkable piece of poetry contains a list of quotations from the Bible, complete with chapter and verse. Alongside these references, which do not quote the actual text, the author has made short comments of a phrase or two. Oddly, his comments do not seem to be relevant to the quoted Biblical texts. This may not seem to be particularly significant to the reader until I point to the fact that there are no less than one hundred and forty-five quotations occupying seven pages of the manuscript! This does not take into account the fact that sometimes the quotations referred to include, for example verses 3 - 11 of Matthew 5, and are therefore involving several verses. This part of the *Apologia* could easily conceal a message, and it is hard to think of any other reason for including so many references, particularly as he does not actually reproduce the text in any of them. It is just possibie, I suppose, that he was showing off his knowledge of the Bible, but there is another odd thing about the quotations. Some of them have no apparent relevance or overt message to convey, so not only are his comments next to the references irrelevant to them — they are irrelevant in themselves to any religious message that he might have been deemed to be sending. So what other purpose could he have had in including them — and so many? Did he really expect people to buy his book and look up every single Biblical reference and read it. I did, and it takes a long time. But, even more mysteriously, four quotations are repeated, which may be an error on his part, but I do not think so. I have a copy of his work which contains handwritten corrections to spelling mistakes and other comparatively trivial errors. It is inconceivable that he would have added the same quotation twice, or more — and not corrected it. It seems to me to be more reasonable to assume that he used those particular Biblical references more than once because they included a word or phrase which he had found impossible to locate elsewhere in the Bible; and that fact may be of assistance in breaking the code, if such it was.

I have considered the many codes available to Orffyreus in the library at Leipzig in the early eighteenth century, and have eventually ruled out most of them, because they would require the presence of nonsense words which would provide too simple a clue that there was a hidden message. Although I am not suggesting that this is the way the message is coded, an example of how simple it could be is this one from the mid seventeenth Century.

Sir John Trevanion, a Royalist, was captured and imprisoned in Colchester Castle. He feared that he would be executed for his Royalist sympathies, as many of his fellow Cavaliers had been earlier. He received a message from a friend which was passed to him without suspicion. The message written in the style of the day, with all its variety of spelling, went thus,

'Worthie Sir John,
Hope, that is ye beste comfort of ye afflicted, cannot much, I fear me, help you now. That I would say to you, is this only: if ever I may be able to requite that I owe you, stand not upon asking me. 'Tis not much that I can do: but what I can do, bee ye verily sure I wille. I knowe that, if dethe comes, if ordinary men fear it, it frights not you, accounting it for a high honour, to have such a rewarde of your loyalty. Pray yet that you may be spared this soe bitter, cup. O fear not that you will grudge any sufferings; only if bie submission you can turn them away, 'tis the part of a wise man. Tell me, an if you can, to do for you anythinge that you wolde have done. The general goes back on Wednesday. Restinge your servant to command.

R.T.

Trevanion read the letter and deciphered the message as, ' Panel at east end of chapel slides'. After visiting the chapel for prayers, he made his escape. The code is very simple — the third letter after each punctuation mark spells the message. I do not think Orffyreus would have used such a system on its own, but he could have incorporated it into another, more complex system. The problem is that there are so many possible alternatives, and so much text to look at. The area most likely to yield positive results, in my opinion, is the list of Biblical references for several reasons. For a start, attention is drawn to Biblical quotations by the inclusion of them at the very beginning and the end of each publication. Next, those quotations are underlined by including capitals which turn out to be Roman numerals giving the date of publication. Add to this the fact that the name Orffyreus is based on the albam method and that draws attention to the other Hebrew code, the number transposition one.

As if the inclusion of 145 quotations was not sufficient to draw our attention to the possibility of there being a secret message hidden somewhere in its text, the *Apologia* has one more puzzle to entertain us. Towards the end of the first part Orffyreus includes the following words, which I have included in full as they are so thought-provoking.

'Those who are keen to ask questions, should ask them of this little book. My deeds will not be revealed prematurely. Should anyone wish to speculate on the truth, let him now ponder on the rich pageantry of words which I now cause to shower down upon him! ...For greed is an evil plant. An anvil receives many blows. A driver drives. A runner runs. The seer sees. The buyer buys. The rain drips down. Snow falls. The shot-gun shoots. The bow twangs. A great herd of fat, lazy, plump horses wanders aimlessly. The flail would rather be with the thresher than the scholar. Children play with heavy clubs among the broken columns. Acrobats and shadow-boxers are as fleet and nimble as the wind. The cunning cat slinks silently along and snatches nice juicy mice. The dog creeps out of his kennel just as far as his chain will stretch. He knows how to please by playing with his little toys and trinkets. He wags his tail, creeps through the hoop and is rewarded with pats on his paws by the stiff fops who watch him. A wheel appears on the scene - is it really a wheel, for it does not have a normal rim. It revolves, but without other wheels inside or outside, and without weights, wind or springs. Seen sideways or full-face it is as bright as a peacock's tail. It turns to the right and the left; it spins around in either direction, laden or empty.'

The author goes on to include references to Alchemy and Astrology. It is one of the most puzzling passages in the book and obviously designed to be so. But it must have been included for a purpose. He introduces the passage with the suggestion that if the reader wishes to know more he should study this book, and consider the words that follow. He says his works will not be revealed prematurely — does this mean that they will be, one day? On the face of it there is nothing to be gained from the above passage, so one is forced to conclude that any truth waiting to be revealed is hidden. It may have been a metaphorical description of the action of his machine. The phrases and words he uses could be jargon, and I have identified some of them as coming from the trade of organ builder, with which we know he was familiar. Strange as the words are, they must have had a point, but without further clues, the message hidden within has evaded me. It may, after all have included a code, and been meaningless jargon to obscure the real message. Perhaps time will tell.

When considering what form the code took, it is worth bearing in mind what information the inventor wished to pass on. If we assume it was instructions for building his machine, then 145 letters does not sound a sufficiently long piece of text for the job. 145 words on the other hand does seem to me to be about right, and in this case there is another element to the puzzle which may help. On the very last page of the document under the biblical quotation referred to earlier, "and do you still not understand?" and containing the Roman capitals, is a simple engraving. It consists of a circle divided into six segments. Three of them are right angled and shaded in — the other three are white and are 30 degrees each. In the middle is a small white dot surrounded by a shaded circle, then a narrow white circle and then another narrow black

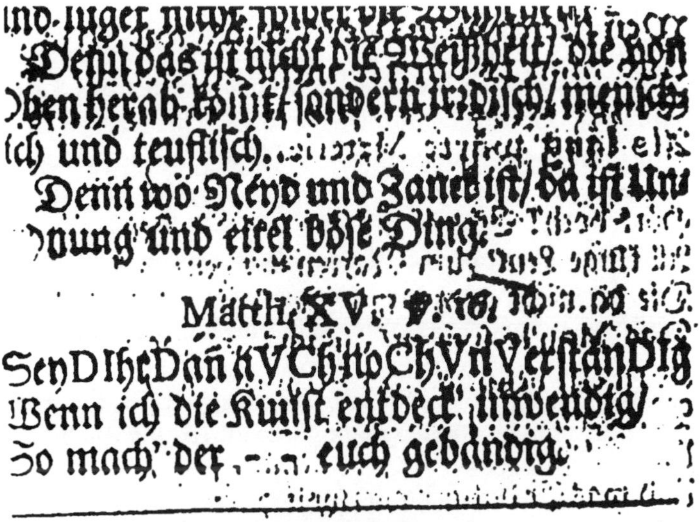

Denn das ist nicht die Weisheit / die von oben herab kommt / sondern irdisch / menschlich und teuflisch.
Denn wo Neyd und Zanck ist / da ist Unordnung und eitel böse Ding.

Matth. XV.

SeyD Ihr Dann AVCh noCh VnVerstanDig
Wenn ich die Kunst entdeckt' inwendig
So mach' der [illegible] euch gebändig.

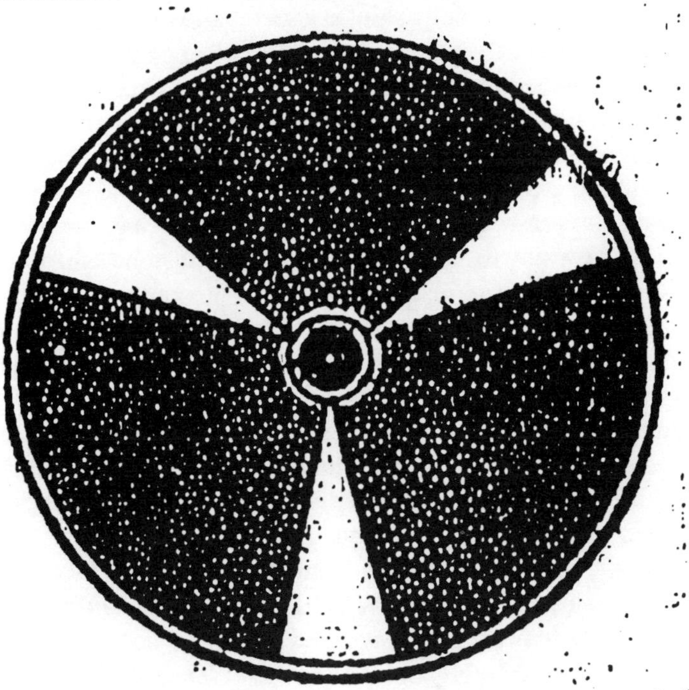

This is the 'wheel' engraving at the end of Bessler's 'Apologia'. Above it can be seen another version of his biblical quotation including the Romanised capitals. The large segments measure approximately 90% and the smaller ones 30%. The centre appears to represent the bearings, then the axle and the wheel itself.

circle. Finally there is another narrow white circle. The whole construction looks like a wheel with a thick axle in the centre. I believe the purpose of this engraving is to enable the inventor to provide instructions for building his machine referring to the illustration as and when necessary.

As for the evidence of a hidden code, I will summarise the facts which lead to me to suspect its existence.

1) Johann Bessler calls himself Orffyreus - Wagner says it is to escape criminal prosecution. I believe an unusual name would be a disadvantage, a common name would make more sense. What other reason could there be? A stage name?
2) The word Orffyreus can be proved to originate via the albam method, from the name Bessler. This method was taught to him by the Rabbi in Prague. He therefore knew about codes.
3) The quotation on the title page of *Apologia* was too vague, there were other more suitable ones. It seemed to hint at a secret known only to the inventor and Karl, which was true.
4) Why use Latin capitals in place of Gothic capitals and lower-case letters in the quotations? Capitals which when regarded as Roman numerals added up to the year of publication, and the year of the construction of the biggest wheel. Was he trying to draw more attention to the quotes?
5) Why did he request permission from Karl to erect a grave in his own back garden? Did he want to construct his own tombstone and therefore have control over his memorial inscription, so that he could give suitable clues to its decipherment?
6) Why did he include 145 Biblical references in his *Apologia*, references that were for the most part irrelevant? Were they coded instructions for building an Orffyrean wheel? Did he really think anyone would take the time to look up the references if they were, as perhaps has been assumed, just included to prove his knowledge of the Bible. In that case a few choice quotes would have served rather than so many vague ones
7) What was the purpose of the very strange passage reproduced above, which he had included in his *Apologia*? Was it just to mystify, or did it contain clues to the secret of how his machine worked?
8) Why include an apparently meaningless and superficial looking engraving at the back of his book unless it was required as part of the instructions?

Orffyreus kept a model of his perpetual motion machine with him through all the years following Karl's death. When he could not sell the device as a perpetual motion machine, because of the malicious rumours of his alleged deception, he continued to try to sell it under the guise of various other contrivances. Automatic fountains, carillons and a submarine were all offered for sale. This does add to the feeling that he believed in his machine, because always at the basis of his designs was the perpetual motion which, if it had not been genuine, would have rendered the latest inventions useless. When he died, the machine was found to be in pieces. He knew his tombstone would provide the necessary information which would lead to the unravelling of the clues, so only his death would, in time allow the secret to become known. With it would go the certain knowledge that he, Orffyreus, had been the discoverer of the secret of perpetual motion, and his fame would be assured. I say this because I believe that what Orffyreus sought was, ultimately a kind of immortal fame, hence his plans to found a new kind of school called the Fortress of God. Unfortunately his immortal fame rested on the actions of his enemies who succeeded in blackening his name to the extent that he was never able to sell the machine, thus depriving future generations of its wonderful benefits.

Knowing of his thirst for fame and fortune, is it likely that he would have gone to his grave without leaving some means of obtaining posthumous recognition? Would he have written on his *Maschinen Tractate* the information that although the actual relevant drawings were missing, someone with an acute mind could, by studying the remaining illustrations arrive at the solution to perpetual motion? The tract is supposed to contain 141 engravings. Intriguingly the final page contains four small engravings which appear to be out of character with the rest of the book. All the other pages contain drawings of theoretical machines, but the last page contains illustrations of children's toys, and the numbers 138,139,140 and 141 have been added in handwriting at the bottom. It is as if having removed the important pages revealing the secret of his machine, the inventor has inserted alternative engravings all on the one page, with the message that study of several different drawings will eventually reveal the secret. This last page may contain a clue to the secret mechanism.

As I said at the beginning of this chapter, the evidence is thin, but it is there nevertheless. A message was left for posterity to find, and I hope that some time in the future, that message will be revealed. I have so far been unable to decipher it, but there is another angle to approach the problem from, and that is this. Orffyreus made a machine which appeared to be a perpetual motion machine. Perpetual motion is impossible, therefore it may have appeared to conform to the requirements of a perpetual motion, but it cannot have been one. How could it do this? Is there any way such a machine could perform as his did without contradicting any scientific laws or beliefs? Could such a machine be made today? Why has one not been made before, if it is possible?

I intend to demonstrate in the next chapter that Orffyreus' machine did not contravene any of today's scientific laws. It was not a perpetual motion within the definition of perpetual motion, but it was what he and most others meant by perpetual motion. It was not the inventor who was wrong — it was the mistaken application of the term perpetual motion to his device, that was the error.

Chapter Twelve

THE SECRET

To get to the truth we must first, as Professor 'sGravesande said, establish what we mean by perpetual motion. Several definitions exist in many dictionaries, encyclopaedias and text books, but the following is as good as any. In fact all the definitions are almost word for word the same, even those in other languages, but those that do differ, say the same thing in a more circuitous and wordy route.

A Perpetual Motion machine is a device which, once having been started, will continue to run indefinitely, with no additional input of energy, as long as the materials of which it is constructed do not wear out.

We may now forget the part about the materials not wearing out, it can be assumed that that proviso is common sense. Perpetual motion of the first kind is the one that concerns us, because it appears to be a machine requiring more than a hundred per cent efficiency. It delivers more energy than it consumes. It is often claimed that such a machine depends on gravity, or at least it would if it was possible. Orffyreus tells us over and over again that *his* machine depends on gravity. We know his machine worked; we know it depended on gravity because it was driven by weights. Let us look at the definition of perpetual motion again, without its assumed proviso, that all the constituent parts should survive in working order.

It is a device which having been started, continues to run indefinitely (for ever!) . . . .with no additional input of energy.

Now it *says* 'NO ADDITIONAL INPUT OF ENERGY — but — if it depends on **gravity**, then **gravity** is providing some input of energy! The inventor admitted that his machine was driven by weights; he even extracted them for Professor Wolff to weigh in his hands — they were estimated as weighing about four pounds each. Fischer von Erlach stated that 'the sound of about eight weights may be heard landing gently on the side toward which the wheel turned'. In his *Triumphans Orffyreanum*, Orffyreus describes the action of his wheel thus,

'Unlike other automata, . . . these weights . . . are the essential parts, and constitute the perpetual motion itself; since from them is received the universal movement which they must exercise so long as they remain out of the centre of gravity; and when they come to be placed together, and are so arranged one against another that they can never obtain equilibrium, or the punctum quietus which they unceasingly seek in their wonderfully speedy flight, one or another of them must apply its weight at right angles to the axis, which in its turn must also move'.

Orffyreus was utilising the energy of gravity. *His* machine was receiving *an input of energy*! To paraphrase Fischer von Erlach, 'at every turn of the wheel there was an additional input of energy from the fall of about eight weights'. Or to put it another way, Orffyreus' machine was a device which having been started, continued to run indefinitely, provided it had the *additional* input of energy from eight falling weights. Of course in this case it could not possibly be a perpetual motion machine, at least not as defined by scientists — and therefore *not* impossible. The strange thing is that many people have always concluded that perpetual motion might be possible if one could somehow utilise the effects of gravity. No one seems to have realised that if one contemplated using gravity, to provide energy, then the resulting machine could not be a perpetual motion machine within the definition. If such a device was not perpetual motion, then neither was Orffyreus' machine. I have said many times that Orffyreus' machine was not a perpetual motion machine but it *was* what he *meant* by the term. All perpetual motionists sought to invent a machine that ran of itself — without requiring such things as wind, water, animals or humans to provide the motive power. Gravity, although the term may have been unfamiliar to them, was a legitimate source of energy to use, simply because it did not need to be added to the machine; it was everywhere and universally available. There was no problem until the scientists decided to scientifically define perpetual motion. When they did that, they automatically excluded the possibility of using gravity, an exterior force, in a perpetual motion machine.

This is all theoretical and I am sure that many will, at this point, argue that it is useless to propose a machine which was driven by falling weights, because those weights have to be raised first in order to be allowed to fall - raised against the force of gravity. This is not such a problem as it appears and I shall come to that in a minute.

So — the definition explicitly states that there must be no additional input of energy; but that is precisely what Orffyreus' machine does have — through

Small bells suspended from a frame, depicted in the upper part of the picture. Two hammers can be seen which were used to strike the bells. Was it the hammer to which Orffyreus referred?

the force of gravity. I have repeated these words because it is my experience that this simple, logical statement, provokes intense exasperation in many well-educated people. For some reason it seems to be a very difficult concept to grasp, perhaps because it has been so deeply inculcated in us, that perpetual motion is impossible. I have taken time on numerous occasions in the past, to explain exactly why Orffyreus' machine *is* possible, only to be met with the standard response — perpetual motion is impossible! I agree -perpetual motion *is* impossible! Orffyreus' machine was gravity driven, and not an example of perpetual motion.

Let us consider the problem of raising the weights in order to allow them to fall. Professor Laithwaite, best known for his work on linear conduction, is also recognised for the excellence of his lectures which used to be broadcast by the BBC. One of his favourite ways of explaining how something works is to use an analogy. As he says, 'You invent a story that sounds as if it could be true, just to try to understand, if only in part, things like electricity and magnetism that would be otherwise quite impossible to handle. So we say electricity is like water flowing in a pipe, and at once we have a feel for electro-motive-force as a driving force or electrical pressure'. So let me offer an analogy.

Consider a Savonius windmill. This idea was invented, or rather reinvented in this century by George Savonius, a Finnish engineer. It was a windmill mounted on a perpendicular axis. The idea is said to have originated in Persia about AD 700, and windmills of this design had the advantage of always being in the right direction to take advantage of the wind. It was the self-same principle that Orffyreus intended to use in his Fürstenberg windmill, when he fell to his death. The Savonius windmill was an ingenious and simple idea. Take a forty gallon drum, and cut it in half along its longest axis. Join the two halves, edge to edge, in such a way that one hollow or concave part is facing in one direction, and the other half's concave part is facing the opposite way. They should be mounted on a vertical axis. Viewed from above the curved surfaces should resemble an 'S'. Place the Savonius windmill in a suitable place with a steady, strong breeze. When the wind presses on the concave face, the resistance to its pressure is greater than that obtained on the convex side. The result is that the side that experiences greater pressure is forced to turn away from the wind, and consequently, the side that experiences less pressure actually moves against the wind. This enables the Savonius windmill to turn in the wind, moving one surface backwards into the wind and allowing it to return to its starting place, over and over again. No one, to my knowledge, has ever suggested that a Savonius windmill will not work because one side of it could not move *against* the wind.

The same argument may be applied to the effect of gravity on a falling weight. Picture a disc or wheel, mounted on a pair of bearings. On each side of the disc are equal weights, attached at a position exactly opposite the other, and at equal distance from the centre of the wheel. The wheel is balanced. Now the force, gravity, is moving in a perpendicular direction downward. In this case the axle is horizontal instead of perpendicular in order to be at a right angle to the approaching force, just as it was in the analogy. The force of gravity is applied to the whole machine just as the whole of the force of the wind is applied to the Savonius windmill. In each case only those parts which can move show any visible effect. Just as the force of the wind was *more* effective on one side of the Savonius windmill than the other — even though it applied that force *equally* to both sides — so the effect of gravity can be made *more* effective on one side of a wheel carrying weights, than the other. How? By simply having one weight nearer to the centre of gravity. I make no claim for this discovery. It is as old as the history of perpetual motion! It is nevertheless a valid proposition and just because it has been included in almost every would-be perpetual motion machine since apples began to fall, does not negate its truth. In other words the *falling* weight weighs the same as the rising weight but because of its position relative to their centre of balance, gravity has a greater apparent effect on it. It acts as if it weighed more than the rising weight. Simply by moving the weight further out from the centre of gravity when it is falling, and moving it closer when it is rising, allows the device to move in a circular direction. This proves, as if we did not already know, that *two weights of equal mass can alternately raise and lower each other* under certain circumstances. A small amount of momentum can be encouraged by making the difference between the positions of the two weights less, thus overcoming a tendency for them to stop in a vertical position. How you arrange for the weights to move in and out is not the issue here, I am merely demonstrating the feasibility of the idea. There is, as I have said, nothing new in this suggestion — it has been offered as a possible solution for hundreds of years and it is so obvious and elementary that it is incredible that science persists in ignoring its inherent possibilities.

Discussions that I have had in the past concerning the above analogy, necessitate that I should point out that where and how the wind originates is irrelevant; the fact that it is initially powered by the action of the sun on our atmosphere does not effect my argument. I am using only the *pressure* of that wind on the windmill to explain the effect of gravity on Orffyreus' machine. It is also worth establishing the fact that my hypothesis does *not* require the consideration of such suggestions as 'gravity waves', 'earthly vibrations', 'the coriolis effect', 'magnets' or other 'gravitic variations' as has been offered to me by various well-meaning people.

We have established that Orffyreus' machine was driven by the power of gravity and it was not a perpetual motion machine. This does not mean the inventor was a liar, it is just that his machine did not comply with the definition. The error, if there is one, lies in the definition of perpetual motion, or rather what can be included under that definition. No one can fault the definition. But it seems to have escaped notice that if a machine relied on gravity for its power it could not be a perpetual motion machine. Otherwise one might be tempted to include grandfather clocks and water wheels! Another way to look at it is to imagine Orffyreus' machine being sent into space. In conditions of zero gravity it is obvious that it would not work. It must therefore rely on gravity. So, it is not a perpetual motion, and *consequently not necessarily impossible!*

Even though we can see that Orffyreus' machine might be possible, we are no nearer knowing how it worked. Apart from the weights is there anything else to be gleaned from the various descriptions available to us, that might give us more information about the actual workings of his machine? In fact there is an enormous amount of circumstantial evidence which points us further in the right direction. The best way to learn more is to study the giant wheel at Kassel.

Orffyreus' own description is as good as anyone's. In *Triumphans Orffyrean*, he describes the wheel as being in the form of a drum, twelve feet wide, with a thickness varying from fifteen to eighteen inches. The axle is six feet long and eight inches thick. It is possible to estimate the weight of the wheel minus its weights and this can help in assessing how much power was inherent. According to a number of sources, before translocation of the wheel, the weights were always removed. This had to be done to allow the wheel to be lifted. Orffyreus answers one of Wagner's criticisms, that this should not have been necessary, by pointing out that,

> 'the empty wheel was so heavy that it could hardly be lifted to its new bearings. With the weights it would have needed the Devil to lift it'.

If the wheel took a great amount of strength to lift when it was empty of the weights, it gives us an idea of its weight. Examining exactly how the translocation of the wheel was effected gives us further information. Two men would be needed to lift it, one on each side, and due to the height of the wheel they could not see each other. They would have to be guided by a third man, probably the inventor. A twelve foot wheel would obviously have an axle at least six feet off the ground, so the men lifting the axle on their shoulders, would need to stand on some kind of sturdy platform on each side of the wheel, but within the wheel's supporting posts. This platform would have to stretch from a point before the centre of the wheel when it was mounted on the first set of bearings to a point beyond the centre of the wheel when it was mounted on the second set of bearings. One witness reported that the wheel was carried five or six steps to the second support. Orffyreus would stand by the bearings to guide the journals into the second set. The reason for the length of the axle being six foot becomes clear. The two men lifting the wheel would need to be between the supporting posts and the wheel in order to lift it and carry it to the second set of bearings without having the supporting posts getting in the way. The platform, itself would have been a necessary part of the equipment anyway, because without it, the many examiners would have been unable to inspect the bearings and journals closely, as they would have been at least six feet off the ground, and probably nearer six and a half feet, to allow a clearance of six inches between the rim of the wheel and the ground.

By estimating the maximum lift likely to have been achievable, we can take a guess at the approximate weight of the empty wheel. It has been suggested to me that a weight of two hundred and fifty pounds to three hundred pounds might be lifted by each man using his shoulders, so let us assume that the wheel when empty weighed around five hundred pounds. Having arrived at a speculative maximum weight for the wheel minus its weights, we need to

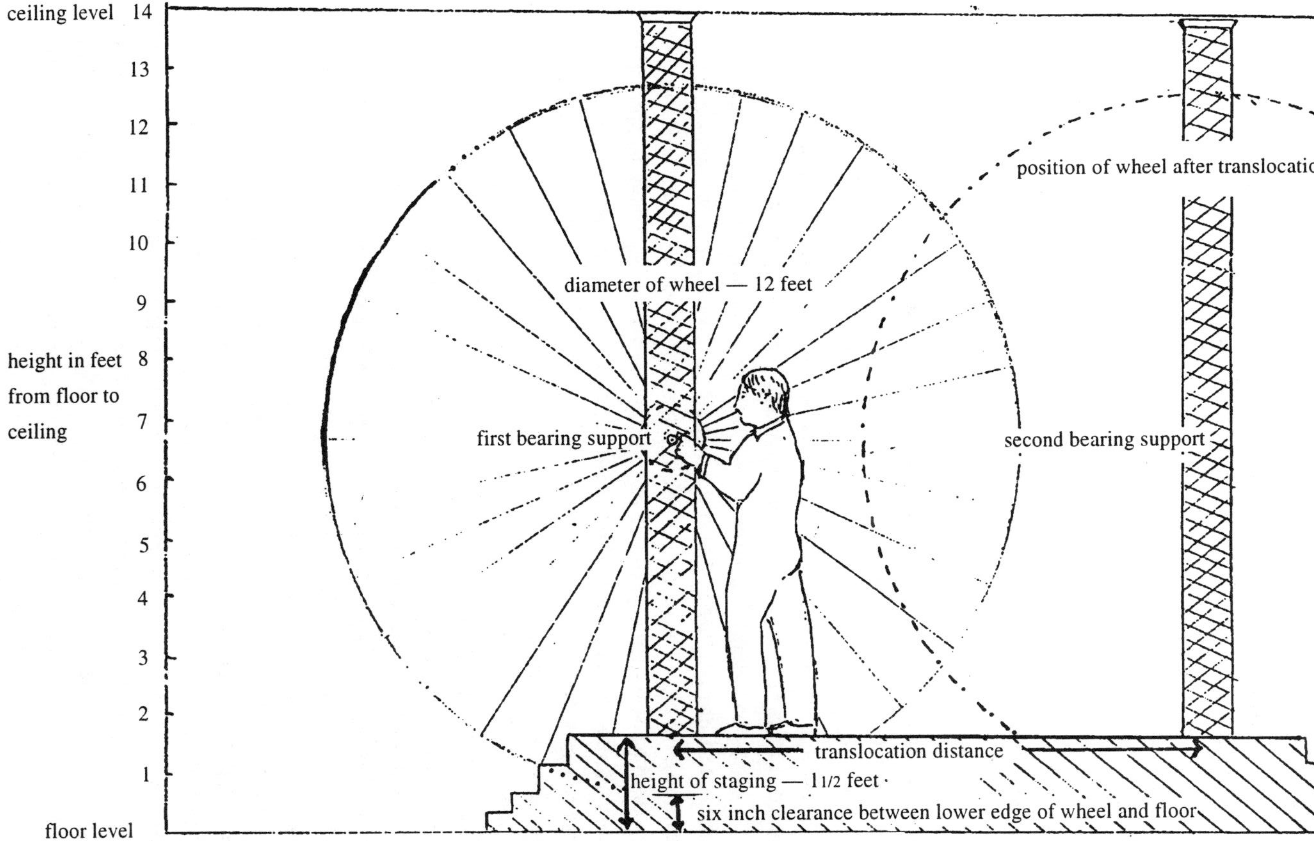

Illustration of staging requirements to allow access to axle of the wheel for purposes of translocating it to another bearing support — and the inspection of the bearings.

estimate its minimum weight. The best way to do this is to design a skeletal framework for a wheel of comparable size, incorporating any features which we know were present in Orffyreus'. Having done that we can work out the volume of wood present and, by attributing that volume to water, whose weight we know, and then multiplying that figure by the specific gravity of, say oak, we can arrive at the weight. Measurements varied enormously at this time so approximations have had to be made, nevertheless, we know the wheel was twelve foot wide and from fifteen to eighteen inches thick, so we shall not be far out at the end. I shall not include all the figures, but the reader can do a similar exercise for his or her satisfaction.

The wheel had eight weights, so I have included eight spokes. Each spoke would need to be joined at its outer end to each other both longitudinally and laterally, and also half way along to provide stability. A glance at the diagram will show how simple the structure is. I have gone for lightness with strength, to try to replicate as light a structure as possible. Using a specific gravity figure for north European oak, I arrived at a figure for the weight of the structure of 300 lbs. The thickness of the spokes or the braces is not known, but I have erred on the side of lightness. I also do not know the weight of the canvas covering of the wheel, but if we assume it is similar to oil-cloth, which one writer described it as, then its weight probably added no more than fifty pounds. Doubling the thickness of the spokes, but not the width, would increase the weight of the wheel to 418 lbs. If the inventor had used pine instead of oak, for instance, he would have been able to reduce the weight, but might have had to use thicker spokes and braces to provide the additional strength missing in pine. In the end he would still have had a machine weighing about 418 lbs. Add the weight of the canvas and the total weight of the wheel minus its weights would have been around 470 lbs. We have already estimated the maximum weight of the wheel as around 500 lbs, so the maximum and minimum weights seem to agree roughly.

There is a lot of information about the weights themselves. Wolff described them in letters to Peter, the Czar of Russia, as being cylindrical, not very thick and estimated to be about four pounds in weight. We know there were a number of them, but not how many. The weights, at four pounds each, must have resembled the head of a four pound hammer. What material they were constructed of is not known, but whether it was lead or iron, the size is not going to vary a great deal. The wheel was between fifteen and eighteen inches thick, and it is hard to know what part of the wheel this refers to. It could be that the outer half of the disc of the wheel, was an inch and a half thicker than the inner half, or vice versa. Or it could be that the struts or spokes of the wheel were visible and were part of the eighteen inch thickness and the spaces between were fifteen inches thick. From this one could assume that the spokes

fig. 1

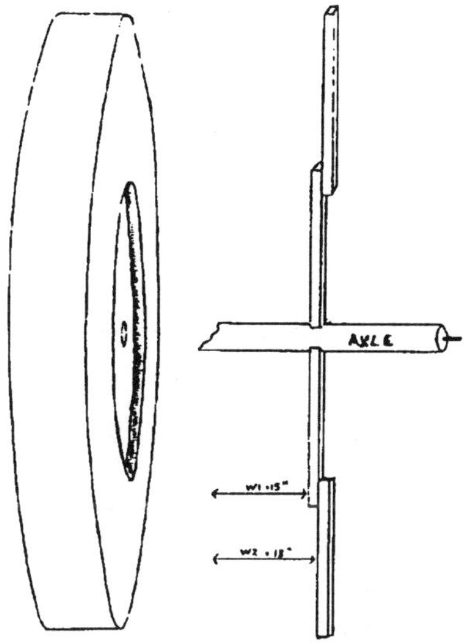

fig. 2

In Fig. 1 there are eight spokes, for the eight weights. They are in pairs, because we know the wheel was eighteen inches wide, and pairs of spokes would be required for stability and strength. I have given each pair of spokes a brace at their outer ends to prevent individual movement, and an inner brace at the halfway point because that is where I believe Orffyreus suspended his weights; he called it a cross-bar. There is, in addition, extra bracing not shown on the illustration for fear of uneccesarily complicating it, which extends from the outer end of each spoke to the outer end of the next spoke, all around the wheel on both sides. This is to prevent the rotational oscillation, and again, to provide stability. This, I believe, is the minimum structure required to support the weights operating inside the wheel.

Fig. 2 shows an edge on view of the wheel which demonstrates why descriptions of the wheel say it varied in thickness from fifteen to eighteen inches. On the right of fig. 2 is a diagram of one spoke projecting from the axle for six feet. It can be seen that the spoke is composed of two pieces of timber joined in the middle by an overlapping joint. The double headed arrow 'W1' shows the interior width of fifteen inches, and the second double headed arrow 'W2' shows the outer part of the wheel has a wider interior width of eighteen inches. We can therefore assume that the spokes are from one and a half inches thick. It is most likely that the weights are suspended from this point via the cross-bar.

were one and a half inches thick, although they could have been very wide. Their width in the longitudinal direction of the wheel was more important than their thickness across the wheel, because it was in the former direction that the weight would be born. If the wheel was eighteen inches thick at the spokes, and fifteen inches thick in between, then the internal thickness of the wheel must have been under fifteen inches. Allowing for spacing for freedom of movement for the weights, this allows room for a maximum of about six weights in a row. This figure is arrived at by assuming that the weights take up about two and a half inches in thickness. If each spoke has a row of six weights somewhere on its length, then, multiplying the six weights by eight, a figure of 192 lbs is arrived at for the total weight of all the weights. Adding that to the previously arrived at weight for the empty wheel of about 500 lbs. we now have the estimated weight for the whole wheel of approximately 700 lbs. Of course there could have been more or less weights and I shall explain in a while how we can narrow down the possibilities. But bearing in mind that the inventor said that his weights acted in pairs, it is just possible that he had another eight rows of six weights, which would bring the total weight of his machine to 900 lbs! Another variation could have been in the number of weights across the thickness of the wheel. I have assumed six, but it was probably fewer. I shall explain why.

It will be remembered that Orffyreus actually exhibited two different types of perpetual motion. In the beginning his wheels were only able to turn in one direction, and also they had to be secured to prevent them moving spontaneously. This suggests that they were in a permanent state of dis-equilibrium. After his Gera and Draschwitz exhibitions he devised a new, improved version of the wheel, which was able to turn in either direction. He says he did this to silence those who suggested that he used a winding mechanism. The second wheel was also different to the first type because it remained stationary once it had been stopped. It did not need to be secured. To start it, it required a gentle push, after which it accelerated to its maximum speed. Orffyreus was right to construct this new version, even though there is evidence that it was not as powerful as the other earlier models. His adversaries had accused him of producing a machine that needed to be wound up; if it could turn in either direction, that would be a difficult accusation to maintain. Wagner did try to cast doubt on the new version too, by publishing news of his 'split axle' wheel, which he claimed would allow the wheel to be turned in either direction. How this was to be achieved he never explained, and the machine never materialised. Orffyreus says it was supposed to be held together with a pin. He, tired of all the libellous comments about him tells Wagner to make a machine that acts as his does . . . or shut up. As he says, if Wagner could do that while admitting that his is a fake, it would be a far stronger argument against Orffyreus.

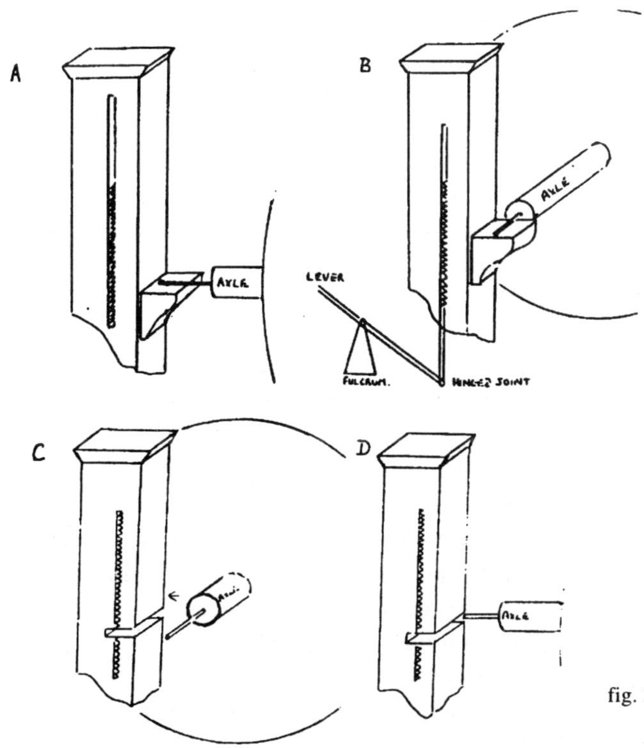

fig. 3

There are only two ways that the Wheel can be mounted on the supporting pillars or posts. The method illustrated in Fig.3. the left hand upper diagram, 'A'. would allow the Wheel to be translocated to a similar post, orientated in the same direction as the first post. In the case of the right hand, upper diagram. 'B', the second post would have to be positioned with its bearing-supports facing the first post. There are advantages and disadvantages in either method, but, clearly the axle. or at least the thin bearing on the end of it, would need to penetrate the post in order to allow the 'barbed iron' to make contact with it. Drawings 'C' and 'D' show how this might be done. This method could have been used, except that the descriptions plainly state that a) the bearings were open to inspection and b) the bearings were uncovered for examination. The latter seems to imply that some kind of cover was removed to allow the inspection to take place; probably the bearing shell, normally kept in place to prevent the ingress of dust or other harmful substances which might increase friction, and wear. Obviously, 'C' and 'D' would not permit either open inspection of the bearings, which would have to be placed inside the slot, nor allow the removal of the bearing cap. I have attempted to show the 'barbed iron in position to show how it was supposed to act on the bearings, and in 'B', I have added a rudimentary lever and fulcrum to try to explain its supposed mode of action. It is self-evident that the power generated by such a mechanism would be insufficient to move such a large wheel.

If the first wheels, the type 'A' as I shall call them, were always out of balance, the second kind, the type 'B' ones, were always in a state of equilibrium — except when in motion. This suggests that when in motion the weights were unbalanced, but as soon as the wheel stopped they regained their balance. If the type 'B' wheels were basically of the same design, but with some modification, as the earlier type 'A' wheels what could have caused this different action?

To answer this we need to put ourselves in Orffyreus' shoes at the time when he was suffering from the accusations that his machine had to be wound up. It was after the Draschwitz wheel had been destroyed. Imagine you are seeking a way around that accusation. You have a sketch of the workings of your type 'A' wheel in front of you and you are thinking, 'how can I make this wheel turn in either direction?' Suddenly you think to yourself, 'what if I repeat the mechanism in the wheel, but install it as if it was to turn the wheel the other way? Will it do that, or will it cancel out the action of the original mechanism?' There would be only one way to find out, and that was to make a model to see what would happen. In the event, putting an additional, mirror-image version of the mechanism in the wheel did have two negative effects; it cancelled out the spontaneous tendency to turn — from now on type 'B' wheels would need a slight push to set them going; and secondly, they turned at about half the speed of the type 'A' wheels.

This tells us quite a lot about the mechanism. The type 'A' wheels had a weight that fell into a position that caused the wheel to rotate to a new position, at which point the next weight fell. In the type 'B' wheels, the original weight and its effect, would have been duplicated, but in the opposite direction on the other side of the wheel, so any reaction would be cancelled causing the wheel to maintain its equilibrium. Once the wheel was in motion, however, a reaction was allowed to manifest itself, and this can best be understood by returning to our analogy of the savonius windmill. A savonius windmill only turns in the same direction in the wind, it can never turn the 'wrong' way. But if *two* savonius structures were constructed on the *same* perpendicular axis, one above the other, and one with its concave and convex halves in opposite directions to its twin, what would happen. The two versions would work against each other and cancel all movement in the wind. *But* — suppose the machine was given a push in one direction, would it come to a stop? No, in fact it gradually accelerates just as the Orffyrean machine did, but only to half of its un-twinned speed. The reason for its slower speed being the resistance provided by the backwards turning *concave* part. It <u>does</u> turn though, because once under way the resistance to turning is less in the direction in which it is turning, than in the other. Interestingly the savonius windmill duplicates the action of the two Orffyrean machines; the twinned ones being stationary until given a gentle push, and the untwinned ones turning spontaneously in the wind. That is why Orffyreus' 'B' type machine performed as it did, and it is totally predictable using the above analogy. 'B' type wheels were inefficient, but necessary to prove the wheel was not clockwork. 'A' type wheels would have been developed because they had more power available for use. 'B' type wheels required a slight push to overcome the friction involved in having two opposing forces operating within the machine. Once that initial resistance was overcome, the gap between resistance to turning on one side compared to that on the other side gradually increased as did the speed of the wheel, until a point came when the weights were falling at maximum speed. At this point the wheel could not go any faster, because of the limitations in the speed of falling objects, and their mass. Discovering the actual arrangement of the weights and how they operated is the problem.

Several times Orffyreus mentions a phrase which literally translated means excess weight. It is this, he claims, that causes his wheel to rotate. Excess weight is an ambiguous term. In the history of perpetual motion it has come to mean the over-balancing of the wheel by an excess of weight on one side of the axle, compared to the other. But it could also refer, in this case, to the effect of the falling weights impacting on the side towards which the wheel turned, the falling side of the wheel. Excess weight could mean the excess force generated by the impact of the falling weights on the falling side of the wheel, compared to either the weight impacting on the rising side, OR the weight required to be lifted on the up-side; and this impact force would increase with the speed of the wheel, while the force to be lifted on the up-side would remain much the same.

Orffyreus remarks that his machine is so well designed that it is not necessary to stop it to carry out repairs or servicing. This is obviously a ridiculous statement if taken at face value. One can imagine the problems of greasing a bearing say, near to the rim of a twelve foot high wheel; one moment you are bending down to the floor; one second later you have to be able to stretch up twelve feet to reach it! This cannot have been what he meant. But if the wheel did indeed have a duplicated mechanism inside it, then one could still use it even if the mechanism for propelling the machine in one direction had to be removed for repair. This explains an otherwise inexplicable comment, and helps to confirm my theory that the machine had two separate mechanism inside each driving the wheel in the opposite direction.

One can obtain some impression of the wheel by picturing it in your own home. It was twelve foot high so one third of it would have been projecting into the upper floor or roof space. It turned at a rate of twenty-six revolutions per minute, or almost once every other second. Many people commented on the loud noise it made, and

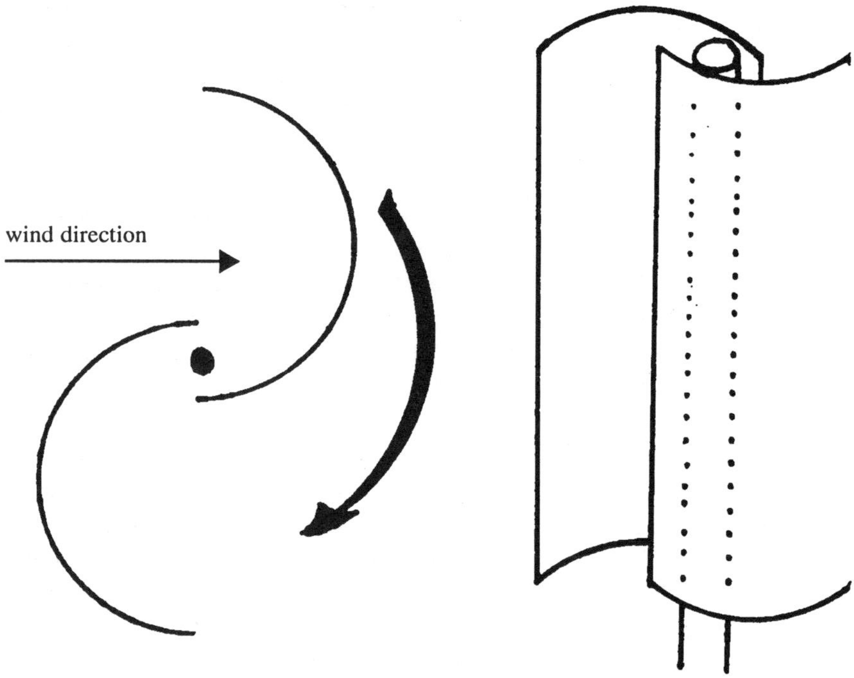

Captain S. I. Savonius was a Finnish engineer, who through exhaustive tests and modifications, arrived at the so-called 'S' Rotor or Savonius windmill. Inexpensive versions could be made by splitting an oil drum in half longtitudinally, and welding the two separate halves together, off-set from the axis, as in the above diagram. Earlier versions of vertical axis windmills are known to have been built in England. But they worked on a slightly different principle. The great advantage was that they worked regardless of wind direction.

as eight weights fell at every turn of the wheel, one can imagine the effect of say a twenty-four pound weight crashing against the side of the machine *four times every second!* The vibrations would shake the house to its foundations.

Having wended our way logically through the evidence to arrive at a skeleton of a theory, I hope the reader will understand why I offer a summary of the arguments put forward in this chapter. In the first place, as Professor 'sGravesande pointed out two hundred and fifty years ago, the only single piece of evidence still in existence against Orffyreus' perpetual motion machine, is the fact that perpetual motion is impossible. I maintain that as his machine did not comply with the definition of perpetual motion, it was not technically a perpetual motion machine, therefore it *was* possible. That being the case there is no reason why his machine could not have been genuine. He thought he had discovered the secret of perpetual motion; it was what everybody meant by perpetual motion; it did not need fire, wind, water or animal power to drive it. In addition I believe that his wheel would have been impossible to fake anyway, particularly in those days.

The evidence of the maid can be discounted for several reasons. Her description of how the twelve foot, seven hundred pound machine was turned by a piece of barbed iron is impossible. Next we are asked to believe that four people turned the wheel day and night, non-stop for fifty-four days, and further consideration lowers that number to almost exclusively Orffyreus' brother, Gottfried. She demonstrated her unreliability by breaking two of the most binding oaths one could imagine swearing, and gossiping to 'mademoiselle', and finally, Karl had seen the inside of the machine — he knew how it worked — and he said she was lying. He was a man of unimpeachable reputation.

We have looked briefly at the lives of those who saw the wheel or were directly involved, and there seems no reason to doubt that they were honest. Karl has been shown to be beyond reproach. How can anyone suggest that he was either party to the deception or gullible enough to be taken in by the inventor? His evidence alone should convince any sceptics that Orffyreus was no fraudster. But we also have the evidence of all those men who examined the machine. Can there be any doubt that all of them would have been searching for signs of fraud, if for no better reason than to protect their own dearly-loved reputations? To be the man who uncovered Orffyreus' fraud would have been almost as big a boost to their egos and their reputations, as discovering the secret themselves. Men like Leibniz and 'sGravesande and Wolff were all convinced that the wheel was genuine, and yet none of them believed it to be true perpetual motion. John Rowley was the most skilful designer and builder of mechanical instruments in the country and possibly the world, then. His knowledge encompassed everything from the biggest and heaviest to the smallest, lightest and most intricate mechanisms known. He was aware of friction and its problems and even if no text-book was available to explain how to overcome such problems, pure experience and an intrinsic and instinctive knowledge and intuition guided him. He showed no hesitation in accepting the Orffyrean wheel at face value. He *knew* with a certainty that cannot be easily understood by a layman, that such a device had to be genuine.

Believing and understanding what today's scientists tell us is partly an act of faith — but so is ignoring the evidence of eyewitnesses such as those discussed above. Why *should* we believe that they were wrong? What makes the scientists of today *more* correct than the actual witnesses of the Orffyrean machine? Those men of the eighteenth century may have been less-well informed than today's scientists but they were as equally capable of deciding whether or not a they were being deluded.

My explanation resolves the mystery. There is no conflict of facts. Gravity is the source of the energy in the wheel. The only remaining problem is to discover the precise arrangement of the weights that provided the impetus for Orffyreus' wheel. When considering the internal mechanism of the wheel, it should be remembered that Karl commented on at least three occasions to the effect that the wheel was very simple, and he could not understand why no one had invented it before. Simplicity is the key. Orffyreus, himself says that perpetual motionists have a tendency to add more and more gears, levers, and weights until the whole thing is jammed solid. The records of the patent office show numerous machines with very sophisticated leverage and springs and rotating gear wheels, all of which merely add to the friction — and the certainty that they will not budge. The successful machine will have an elegant simplicity about it, an arrangement that will make people wonder why they did not think of it themselves. Orffyreus himself, comments that when the secret is revealed, he is afraid that people will complain that the idea is so simple it is not worth the asking price.

Orffyreus was determined that no one should rest the secret from him during his life, and although he asked a fortune for the machine, it was not an unheard of price. The British Board of Longitude was set up to encourage people to devise a means of establishing a ship's position at sea. The prize on offer was £20,000 for the inventor who could come up with a solution; the same figure that Orffyreus was asking for. Karl gave away to his mistresses considerably more than that figure. John Harris who finally won the British Board of Longitude prize had great difficulty in actually receiving it. It took him more than twenty years to get it. Orffyreus must have known, towards the end, that his chances of getting payment for his machine, were vanishingly slim. It is inconceivable, knowing him as we do, that he would not have planned to have his secret revealed following his death; revealed in such a way that there was no doubt that he was the inventor. He disassembled his model for two reasons; one was to prevent a burglar or chance visitor from obtaining his secret, and the second was that if he should die, a possibility that he discussed in his last letter, no one would be able to get their hands on the machine and claim they had discovered it.

He was a difficult man, single-minded, self-opinionated, and sanctimonious. He was also devout, ingenious and a tough negotiator. He sided with the poor and the weak and the peasant-labourers, giving help and money to them for many years, but at the same time he cultivated an acquaintance with the rich and noble. He belonged nowhere and yet everywhere. He was capable of displays of great affection, friendly, emotional and highly sensitive — a complex, and painfully vulnerable man, who could at the same time release a torrent of abuse on the object of his disdain. He demanded loyalty and secured it, unnecessarily, with oath-taking of the most extreme kind. Opinion about him was divided, but everyone had an opinion.

Controversial as he was, his claims should not be ignored. The evidence that his machine was genuine is overwhelming, only the single fact that it is supposed to be impossible, stands in the way of accepting its viability. That fact, is a myth. It has been exploded, and it is now time to review the evidence presented in this book and start to investigate the feasibility of producing a machine which, like the Orffyrean machine, derives its power from the force of gravity.

When the day comes that a prototype machine has been developed and the engineers and scientists involved are looking for a name to call their new force, what better name than 'Orffyrean' energy could they look for. It would be a belated recognition for the man who spent his early life looking for the answer, and the rest of it trying to persuade the doubters that he had succeeded. I sometimes imagine him, watching the development of the steam engine, the combustion engine, space-rockets, nuclear power and all the alternative forms of energy; frustration gnawing at him, the pain of over two hundred and fifty years of infamy, burning into his soul, because scientists would not accept the evidence of their eyes; tortured by their disbelief, anger inflamed beyond bearing. How it would salve his silent fury, and ease his bitter memories if at last his true worth could be acknowledged. Does he now, still tremble with righteous rage, spitting violent curses and calling down on those who oppose him, Hell's most evil demons to imprison them for eternity? Recognition of his genius could exorcise his ire, relieve the pain of two and a half centuries of odium, and fulfil his desire for fame and glory.

Chapter Thirteen

PRIME MOVER FOR THE MILLENNIUM?

If Orffyreus' secret was discovered now, in time for the Millennium, of what use would it be? Medieval technology would appear to be out of place on the brink of the twenty-first Century. The original machine was built largely of wood, except for the brass or steel bearings and the metal weights. In modern times an Orffyrean machine would be made of steel or perhaps aluminium. Some smaller varieties could be manufactured out of plastic. The percentage of heavy material would be whittled down to the barest minimum, unlike Orffyreus' wheel which was designed to hide the mechanism, and therefore used an excessive amount of wood. Such a machine could be made even more simple in construction than the original one. Of course today's bearings bear no resemblance to the kind of haphazard fabrication available to Orffyreus. Virtually friction-less components could be built into a modern version of the wheel, ensuring a long and maintenance free life for the machine. But what would be the point? Is there a need for a self-moving, non-polluting, non-resource-consuming, generator? Of course there is! In fact there has never been a better time for its rebirth. Pollution from the burning of fossil fuels threatens our lives; and any way the reserves of those fossil fuels available to us are increasingly scarce. But before we look into the areas of possible development and usage of an Orffyrean machine, I have a suggestion to make.

Orffyreus thought he had discovered the secret of perpetual motion, but in reality his claims had fallen short of the actual definition. He called his machine a perpetual motion or a perpetuum mobile — what if we were to take the first three letters of 'perpetual' . . . and the first two letters of 'motion' . . . and coin the word . . . PERMO? This would contain some of the ingredients of the words in perpetual motion, hinting at it but not spelling it in its entirety.

It would be a shortened version, to convey the fact that his machine fell short of the scientific definition. As I have said repeatedly throughout this book, Orffyreus' machine was not a perpetual motion within the meaning of the definition but it was what he meant by the phrase. In recognition of this I think it would be a good idea to have a word that conveyed that meaning without wrongly calling it a perpetual motion machine. At first sight this may appear slightly ironical, as if somehow his machine was slightly inferior to the real thing, but nothing could be farther from the truth, and when we look at the symbolism contained in this simple word, I believe it will be accepted. For a start take the word PERM in PERMO. It conveys a sense of permanence and durability, as in the endless running of a perpetual motion machine. The MO part could be short for MOTOR, and taken together PERMO represents the everlasting nature of the motor. **O** in PERMO stands for the initial letter of the inventor's name, Orffyreus. During research into this book both I and researchers I have contacted have constantly referred to Orffyreus as O, for the sake of brevity. **O** can also stand for ZERO, or the amount of energy required to be supplied by man, in one form or another to enable the wheel to run. It needs none, as the only energy required is that freely available in the form of gravity. **O** can also stand for the wheel itself, especially if a dot is placed in the middle of it. Orffyreus, himself used to sign his name with the **O** clearly displayed and with a dot in the middle. This would again symbolise the first wheel invented by Orffyreus which displayed his Orffyrean energy. If that is not enough symbolism, one can take the P to stand for permanent or perpetual; the E for energy; the R for resource or renewable, and the MO for motor, or Orffyrean motor — Permanently Renewable Orffyrean Energy Motor. I have included the last because people seem to like acronyms.

PERMO is the name I have coined to describe an Orffyrean machine. As I said at the beginning of this book, I suggested that the simplest and most obvious use would be to attach a PERMO to an electricity generator. This arrangement could be used to power domestic electricity requirements. Instead of taking power from one of the public utilities supplying electricity, one could have one's own independent supply. The first advantage of this would be that one would no longer have to pay electricity bills. Of course there would be the initial expense of buying a PERMO and generator, and afterwards a maintenance cost, however the average American family spends over $1000 per year on energy in the home, and I know that in the UK the cost is somewhat higher. That being so it would not take long to recoup the original outlay, even if it cost say $5000. The machine is so simple that it could be made to run for twenty years or more with little more than the occasional bearing replacement. The generator, being of a more complicated nature might cost more in maintenance, but it is still a relatively simple device, and the savings possible against the cost of using electricity from a power station are enormous. Once installed a PERMO would obviate the necessity for a gas supply. Why pay for the use of natural gas, or oil, when there is a supply of free electricity at hand? The demand for electrical heating appliances would climb dramatically, just as the need for gas operated systems would reduce. Share values in manufacturers of portable generators and electric appliances would soar, whereas shares in gas appliances would stagnate or fall slightly. I say stagnate or fall slightly, because none of this will occur overnight, and by the time that gas appliance sales have fallen to an unacceptable level contingency plans will have been brought into operation, perhaps converting to electric appliance manufacture.

Reaction to a PERMO operated heating or air-conditioning system would be dramatic and probably panic driven, by the major power companies. They would do all in their power to remove the threat to their livelihood, using any methods to achieve their aim. The mildest might consist of buying the rights to such an invention and burying it. That would be a disastrous thing to happen, and it is imperative that such a fate be avoided. In fact there are many good reasons for them not to view the development of a PERMO with alarm. For a start it would take years for every household to convert their electricity supply to a self-supply situation. It would make sense for the various public utilities to get involved in supplying such machines to the general public. They might even lease them, and the profits could be very good. The electric and gas supply companies would still be required to supply industry as well as those millions of people who will never get around to converting to a PERMO. In addition the use of PERMO machines would help to conserve the reserves of coal and gas currently dwindling at an alarming rate.

Electricity is produced by a number of different methods. All of them produce CO2 which is a greenhouse gas and contributes to global warming. Burning coal produces the most CO2 and yet current coal reserves are not expected to last more than another two hundred and fifty years. That may sound a long time, but coal will not run out overnight, it will gradually become more and more difficult and expensive to mine as that day approaches. As reserves dwindle action to find an alternative form of energy will become more urgent. During that time global warming is expected to raise the mean annual temperature by up to nine degrees Fahrenheit, enough to allow the melting of the ice at the poles to lift sea-levels by three feet, drowning thousands of islands and millions of square miles of inhabited lowlands. If global warming was all there was to worry about it would be bad enough, but burning coal produces other harmful by-products which cause more damaging environmental problems such as city smog and acid rain. Sulphur dioxide is pumped into the air and returns to the earth in the rain, causing damage to crops and forests. Even if the coal reserves were plentiful, the search for alternative methods of generating electricity would still be paramount.

It is not just the coal-burning power stations that contribute to global-warming. Transportation on the whole uses petroleum, and this contributes more CO2 to the ever increasing amounts being released into the air we breathe. Another use for PERMO is therefore, potentially, the car. A powerful but compact PERMO could be designed to run a car, and think of the benefits. No need to buy costly petrol. No need for all those expensive garage bills for servicing the very complicated auto engine. Cars would be so much cheaper to make with only a simple PERMO attached to an electric motor. Orffyreus designed his wheel as a tall narrow mechanism because, I believe, he wanted to make it as narrow as he could to make it seem more impressive that so much power was packed into such a limited space. In fact I am sure that a machine could be developed that was more of a squat cylindrical shape, and thus suitable for installation in an auto. The inventor often said that his machines could be run in series, so imagine a machine only twenty-four inches in diameter, and three inches thick. That is a machine of similar proportions to the Orffyrean versions. But instead of putting perhaps half a dozen such machines one behind the other in series, an improved version could probably designed which incorporated the power of all six machines in one unit, eighteen inches in depth. Considerable weight could be incorporated in such a machine, providing a powerful source of energy to drive an electric motor, and thus an electric vehicle. It is inevitable that the huge petroleum producing companies would also react unfavourably at first to news of the PERMO. But again they need have no fear. The reserves of crude oil are much more limited than coal. In fifty years it will become uneconomical to extract oil and then alternative sources of energy will *have* to be found, and Orffyrean energy is the answer. It will eke out the reserves of crude oil and of course the many other by-products of oil will still be required. PERMOs could be built to run lorries, ships and trains, all of them non-polluting and cheap to run. I am sure that as the technology advances, the speed and load carrying capacities will improve dramatically.

Among other forms of energy generation the nuclear industry is a major player. This is a resource which is under attack by environmentalists because of the problems of nuclear waste. No matter what solutions the industry comes up with regarding the disposal of the waste, there seems to be an instinctive dissatisfaction with the proposals. Nobody likes the idea of burying the waste either underground or in the sea, however effective the containers are. Only neutralization of the harmful radiation emitted by the waste would be acceptable to the majority of people, and that is not possible. The poisonous emissions from coal-burning are almost totally absent in nuclear power, but occasional accidental releases of radioactive gases and coolant water are a cause for concern, as is the potential for a more serious accident such as happened with the Chernobyl reactor. Looking at the overall picture, nuclear power may be a useful way of generating electricity, but at what cost? The actual cost of first commissioning and then de-commissioning a nuclear facility is so expensive that it has been estimated that it would save billions of pounds if all the money which had been spent on the nuclear reactors had been spent cleaning all emissions from coal-burning power-stations, so that nothing but pure air was released! This seems to be politically unacceptable. Personally I believe that nuclear power generation has to go, and now that there is the possibility of producing electricity cheaply, without pollution, there is no need to wait any longer. All nuclear plant proposals should be cancelled and a start made on de-commissioning existing ones.

Renewable energy sources are actively being pursued. Hydroelectric schemes are the most successful and countries such as Sweden who already have many hydro schemes in operation are phasing out nuclear power altogether within the next ten years or so. Unfortunately not many countries are suitable for the installation of hydro-electric power schemes for generating electricity. Alternative forms of energy generation include wind-energy and solar energy. The USA have spent millions of dollars on wind energy and one site in California has no fewer than 7,500 windmills. The State has so many that 1% of all its electricity is produced this way. Even so 1% is barely scratching at the surface. Imagine how many windmills it would take to provide enough energy for the world's requirements! Solitary windmills in isolated positions can sometimes look attractively unworldly, but one is enough. Wind farms on the scale of the one in California would be needed in vast quantities to make a worthwhile contribution to public electricity consumption. The other major source of energy under investigation at the present time is Solar energy. This is collected in two different ways. One way is for each site to store the heat of the sun during the day and release it slowly at night. Alternatively photo voltaic cells which convert sunlight directly into electricity are being used. The astronomical costs of these PV cells has dropped and is dropping all the time and eventually may fall to a thousandth of their original cost. There is great potential in this industry, but again it is not a suitable application for the cooler climates where the sun's light is dissipated considerably. The remaining alternative fuels are in the minority. Some employ biological matter that under the right conditions produces methane, for home heating. Brazil produces ethanol by fermenting grain, and this is widely used in cars in place of petroleum. Of course this still produces undesirable by-products in exhaust emissions.

Geothermal energy uses the earth's heat, but it is very limited and is not technically a renewable source. Hot water rising from the depths of the earth is used, but it has been noted that some of the older systems still in use have seen a reduction in available heat energy and therefore can only be regarded as a temporary energy source. Tidal energy machines have been and are being developed but they are unsightly and can restrict access to the sea. Their output is small per unit and they are not a viable option at present. In fact they are not a new idea for supplying energy; one only has to remember that John Rowley designed and built a new tidal wheel to pump water to Windsor Castle, from the river Thames, nearly three hundred years ago.

The only remaining source of energy for transport is the battery, used in the electric vehicle. Many of the world's auto manufacturers have produced from time to time their versions of electric vehicles. Their range is limited by the size of their batteries. Electricity storage can only be in batteries and they are large and heavy. They also deteriorate quickly and are expensive to replace. An additional factor often disregarded is that the electricity they use has to be produced by one of the methods outlined above and they are therefore indirectly contributing to pollution, even though it comes from the power stations rather than from the batteries. Some motor manufacturers have developed hybrid engines — part electric motor and part petroleum, and either system engages as required depending on speed and load requirements. This is ingenious but is still perceived as an unsatisfactory solution to the problem. The resultant engine is more complicated than the standard petrol engine and therefore more prone to breakdown.

The above options open to us at the arrival of the millennium show that there is a good awareness of the problem and great efforts have been made to find an economic answer, but they nevertheless make gloomy reading. The increasing demand for more energy by a growing world population means more pollution if that energy is produced by traditional methods. Research continues in an effort to find a safe renewable source for energy generation. So far no satisfactory solution has been found. But maybe it has, at least in theory. A possible answer lies in Orffyrean energy. A PERMO machine would provide the perfect answer to all of the problems outlined above. There would be no need for coal-burning power stations, and thus no CO_2 and Sulphur dioxide emissions. Acid rain would be reduced, but not eradicated because Steel manufacture requires the use of coking coal which still emits sulphurous gas. There would be no need for nuclear powered electricity generators, so no nuclear waste to process or dispose of. No need to power transport with petroleum combustion engines, so a further reduction in CO_2 emissions. Electricity would be plentiful and cheap. All homes could be warm, and the fear that many of the poor and the old suffer from — over-using electricity or gas and being unable to meet the ensuing bills, or alternatively under-use it and die of hypothermia — would be removed permanently from their lives.

Of course as I said earlier some power utilities might continue to supply electricity to industry. Large users of electricity might find that drawing power from a large supplier such as a traditional power station might be more convenient and cheaper than installing their own large PERMO. Efforts could be made to render emissions from coal-burning electricity supply companies completely free of any kind of toxic substances. It would be expensive but not as expensive as running a nuclear reactor, and the benefits of removing all nuclear reactors from the electricity supply would be a tremendous bonus to cleaning the planet earth.

Running a home on electricity produced by a PERMO, could occasionally throw up a problem. Lighting and most appliances would not place too much of a demand on a home generators, but items such as heating and cookers might pose more of a problem. To level the variability of the demand on the home-based PERMOs, individual appliances could be fitted with the devices. Refrigerators and freezers could be run from small integral PERMOs. Radios, TVs, vacuum cleaners, kettles, microwave cookers, and lamps could all have their own individual power-packs. Each power-pack would consist of a PERMO and a small electrical generator. Lawnmowers, pumps and motorcycles could be designed with similar units fitted plus a small electric motor. Toys for all ages of children could be designed to incorporate PERMOs and one can imagine that this may cause a fall in battery sales, but those manufacturers could sell more rechargeable batteries and also chargers, operated by PERMOs. It is conceivable that standard power-packs, each designed to produce a certain output for particular requirements, could be manufactured for fitting to a variety of finished products.

The era of the 'wireless' electrical appliance is just around the corner — and not just 'wireless' but cost-free running. The potential uses for such a machine are so diverse that it is impossible to describe all of them. Third world and other under-developed countries could make use of automatic water purifiers, refrigerated rooms, irrigation plants, tractors, radio receivers, buses, and portable power generators for hospitals, kitchens and lighting, refrigerators, cookers and various kinds of transport. New ideas will surface from time to time, but what is needed now is concentrated research into the possibilities suggested by this book. Whoever discovers the secret of Orffyreus' PERMO, will have a highly significant effect on mankind's future on earth. That is a dramatic claim but not far from the truth. They will also become very famous and very wealthy — surely there must be someone who would like to enjoy the benefits of discovering the secret of Orffyreus' wheel?

How could such a machine be brought to the public's attention while at the same time protecting it from being monopolized by a major industrial concern? I believe that even if one organisation bought the complete rights to constructing and marketing a PERMO, it would not remain its sole property for long. One of the attractions of this machine is its simplicity. Even though I have not seen the internal workings, I know what Karl thought of the design — he could not believe that no one had thought of it before. He said it was so simple a carpenter's boy could make it, if he was allowed to study it for five minutes. If it is so simple then even if a company owned the rights to it, that would not prevent thousands of people constructing their own machines. Articles would be published explaining how to build personal PERMOs, and what kind of generator to add on, depending on your requirements. Only those manufacturers requiring a high tech approach, such as the automobile industry, would have a kind of protection because they would be developing electric motors most suitable for being run off generators driven by PERMOs, but which could be embodied within an automobile. Theoretically, the drive could be taken directly off the PERMO, but I think that it would be more likely that they would want to generate electricity first and then connect the drive via an electric motor.

Public knowledge about PERMOs would be widespread, and so the most fruitful method of proliferation would be if governments were to encourage independent research and development, by small companies. Each could choose their area of operation; some building home electrical supply units; others moving into the home appliance market; while others looked at the small transport market, powered bicycles and scooters etc. The one essential is that this industry is not in the hands of a few, and I do not see how it could be kept there, even if that is how it began. All those jobs that would gradually disappear from the various power-producing companies, could be replaced very quickly by a large band of entrepreneurs. They would be assisted by those others who wanted a more stable working environment, working for someone else rather than on their own. I actually believe that rather than putting many people out of jobs, this new technology would create far more jobs than it lost. If you just consider the number of dwellings which would gain from the introduction of a PERMO into their running . . . *worldwide*! There will not be enough people in the workplace to cope with the demand.

There is one more facet to Orffyrean energy that may well outrun all the advantages so far discussed. It is purely speculative, but as I have said before, logical. I shall use an analogy again to explain my reasoning. If you pass a current of electricity through the copper windings in an electric motor, the motor starts to turn. Alternatively, if you reverse the process and you turn an electric motor, an electric current is produced in the wires. The two devices are known as a dynamo and an electric motor; two sides of the same coin. I believe that if instead of allowing gravity to drive a PERMO, the situation was reversed as in the above analogy and a small motor was introduced to drive the PERMO, the weights which would normally fall, *could* be designed in such a way that they rose upwards. If a PERMO with motor attached was placed on a set of scales, and set in motion, the weight registered by the scales would indicate a gradual lightening of the load as the speed of the motor increased. The result would be a decrease in weight. An anti-gravity machine! I think it might be possible to design a machine that used this technique to lift itself into the air, and taking the next logical step, perhaps some kind of aerial device could be designed. A good name for such a device would be bootstrap! It is but a short step to envisage an aircraft of some sort. Of course it would not be like a normal winged craft — there would be no inherent lift such as is exhibited by an aerofoil. It would have more of the ingredients of a helicopter, but even this has a wing. A PERMO based aircraft would depend totally on the running of the machine to provide constant lift in order to remain in the air. Following on from this speculation, it is obvious that a ship designed to go into space might be a real possibility. If this is true then it might prove be a much cheaper way to get into space, and once there drive a spaceship in any direction desired. Even in zero gravity it may be possible to use this technique to traverse the reaches of space.

This is all highly speculative and until a proper programme of research has been established, we shall not know if the possibilities discussed above can be turned into a reality. One thing I am certain of is that there will be a reaction of a negative sort by certain members of the public. It is inevitable that some will suggest that no energy source is free; that we will have to pay in the long run for free energy. I do not anticipate any problem dealing with such predictions of gloom. If concern is apparent about the generation of heat to excess, I would point to the fact that we are already producing excess heat via the coal-burning electricity supply companies. That source of heat would be removed to a very large extent. World production of heat would not vary by very much, so the overall total should not rise at all, even if populations continue to rise. A more ridiculous suggestion has been put to me that using the force of a falling weight to produce energy would somehow remove that unit of gravity from the equation for ever! What do they think would happen?

Will we somehow all become weightless and fly off into space? One might as well suggest that using wind power might somehow remove wind from the earth's surface, or using solar energy would actually dim the light from the sun. There will always be those who wish to cast doubt on the benefits of such a system but I cannot imagine that there are any problems associated with PERMOs which cannot be anticipated and catered for.

It has been remarked to me on more than one occasion, that if Johann Bessler discovered the secret of perpetual motion, how come nobody ever discovered it before? Even Karl the Landgrave, expressed surprise at this. We have seen that there may have been candidates for that honour previously, but the question did get me thinking and I offer my conclusions and findings accordingly.

The subject matter in this book concerns events that happened more than two hundred and fifty years ago. But the search for the secret of perpetual motion goes back much further than that. If Johann Bessler could solve the problem with a simple machine in 1712, there seems no reason why a human living two or ten thousand years ago could not do so too. The needs for such a machine would have been the same then as now, it is merely a matter of degree.

For proof of his mental and physical skills one need only look at the surviving edifices that are around us today, built by man thousands of years ago. Take Stonehenge for example; Reputed to have been built four or five thousand years ago, it consists of some eighty pillars weighing about four tons each. The stone for these was transported some 250 miles to their present site. Next some thirty Sarcen stones weighing in excess of fifty tons each were raised. Then, as if this was too simple the builders capped these stones by putting lintels on them, not just by placing them in position, but by actually cutting mortise and tenon joints into each end and the top of the uprights. The icing on the cake was the fact that they made the lintels into a curve in order to conform to the circumference of the circle they described.

The pyramids offer similar impressive references to early man's ingenuity. All over the world, examples challenge modern man's capacity to accept that thousand's of years ago our ancestors were capable of lifting and precisely placing huge pieces of stone, some weighing hundreds of tons. It does not seem very remarkable then, that some of those same people might have been able to design and build a machine similar to Bessler's.

Proof of the existence of such an early perpetual motion machine will be difficult to find. Examination of ancient Egyptian hiroglyphics might be informative. I have found indicators in certain areas that seem to imply the regular use of such machines, and one avenue for exploration lies in symbols.

It seems to me that if such a machine existed in the past, a representation of it might linger on in artwork and design. A simple example is the swastika. It has evil connotations today, thanks to its use by the Nazi Germany, but it wasn't always so. In 1910 a poet and nationalist ideologist, Guido von List, had suggested the swastika as a symbol for all anti-semitic organisations. In 1920, when the Nationalist Party was formed, the swastika was adopted as its symbol, and subsequently became part of the National flag. But in ancient times the swastika was a symbol of prosperity and good fortune and it was widely distibuted about the world.

The word 'swastika' is derived from the Sanskrit 'svastika', meaning 'conducive to well-being'. It was a favourite symbol on ancient Mesopotamian coinage, and it appeared in early Christian and Byzantine art. It was also known in South and Central America among the Mayas, and in North America, among the Navajos Indians. It is still widely used in India and a clear distinction is made between the left hand cross and the right one. The right handed one, moving in a clockwise direction is said to represent the Sun, whilst the other, left handed one stands for the moon.

The swastika lies behind the symbolism of all of the world's many crosses. Under various names such as cross, cruz, crux, crowz, krois and krouz lies an original swastika. The multitude of meanings assigned to this symbol all owe their origin, via the Sun symbol, to the concept of power, energy and life. It would appear to have represented an energy source to the ancients. We have seen that it is distributed widely and it also appears in almost every primitive cult throughout the world — in Ireland and Mycenae — among the Etruscans, the Hindus, the Celts — in Cyprus and Troy — in Iceland, Lapland, Finland, Scotland and England.

Scholars are undecided as to the exact significance of the swastika. It has been described as a revolving sunwheel; the sun chariot; the four cardinal points; the four winds; the four seasons; the motion of the revolving world. It is clear that no one knows and all are guessing, as no written evidence exists to explain how it came about and why it is so widely destributed. More esoteric explanations have been offered. It represents the union of the male and female principles; the dynamic and the static; mobility and immobility; harmony and balance; centrifugal and centripetal, and so on and so on. The device is found on the walls of caves and buildings; alters, figures, urns, vases, utensils, pottery, weapons, shields, dresses, coins and spindle-whorls. In all cases it is said to be a symbol of good luck, good wishes, blessings, longevity and long life.

In summary it would seem that the swastika is a ubiquitous sign that has been around since the dawn of mankind. It is generally asociated with good

26

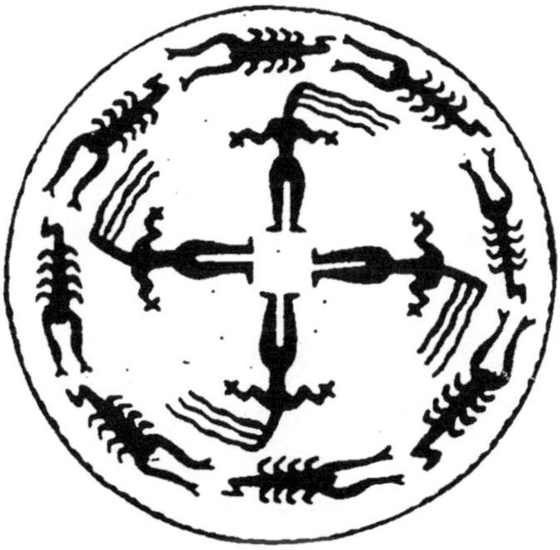

This is a copy of an ancient swastica found on a piece of Summarian pottery. It is formed by four women and their hair, and is said to represent the female generative force. This piece of pottery is at least seven thousand years old.

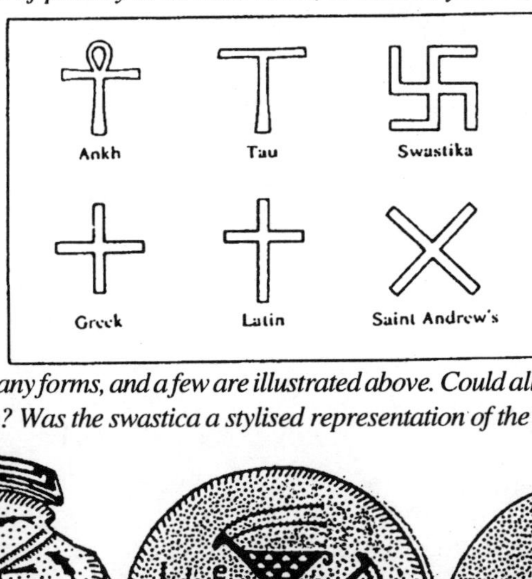

The cross takes many forms, and a few are illustrated above. Could all of them have originated from the swastica? Was the swastica a stylised representation of the Orffyrues machine?

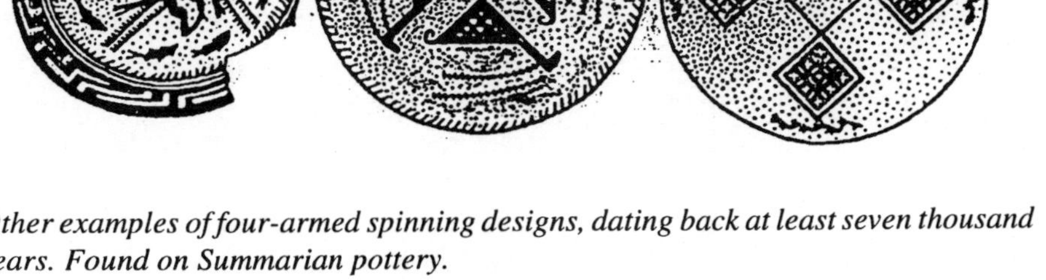

Other examples of four-armed spinning designs, dating back at least seven thousand years. Found on Summarian pottery.

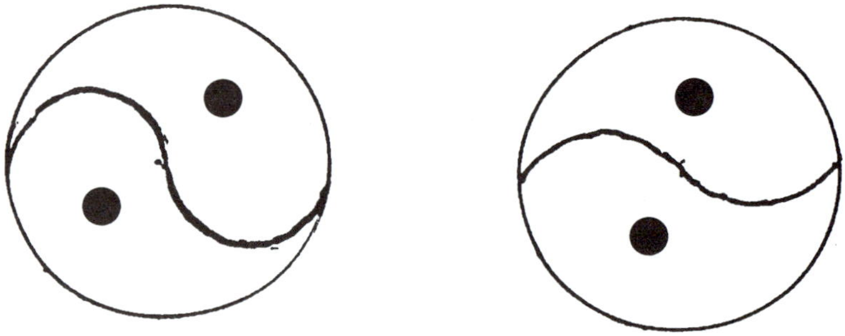

The yin-yang symbol is well-known but little about its origin and purpose is known. The version on the left is the common conception but the figure on the right is closer to the original design, and resembles the plan view of the Savonius windmill.

fortune, longevity and energy. All versions contain the same basic plan. Four arms radiating from a centre; each arm bent through a right angle in the same direction. Why the same symbol the world over? Could this symbol be a stylized version of the plan of Orffyreus' wheel? The wheel that Orffyreus made was said to have eight weights, but his turned in either direction — maybe only four were required for the early machines. One can imagine that a Perpetual Motion machine such as Johann Bessler built would have been built of wood and stone plus a small amount of metal, thousands of years ago. Nothing would have survived to this century if the technology behind it was lost, but it is possible that the basic design might have survived and gathered to itself additional symbolism such as is associated with the swastika.

One other symbol which is also extremely ancient and which is still in vogue today is the yin-yang symbol. I have researched the history of the symbol and found no convincing explanation for its existence, which is odd considering how familiar the sign is. Even the famous Chinese oracle system of I Ching mentions it and describes how it represents the dual forces of opposites; dark and light, hotand cold, male and female, even and odd etc., etc. but it does little else to explain the presence of such an interesting design. One famous expert on China described it as two tadpoles head to head. And this is what I find curious about the design. One is tempted when drawing an accurate representation of the symbol from memory, to draw the two semi-circles as perfect half-circles within the outer circle. But this is wrong. The oldest versions of the yin-yang symbol all have the central curved line as an almost elongated 's' shape rather than two half circles and I think this is a clue to what is really behind the design — one is reminded of the plan view of the savonius wheel. Before I enlarge on that though it is worth considering what the scholars and historians have to say about the symbol.

We have seen what philosophies have been extracted from the symbol, but in addition, they say that the dark half is the heavy part and the white part is the light part. The black portion is always trying to drag the circle down on its side, and white area is light and trying to lift the circle upwards. Does this not sound like an Orffyrean wheel. As well as the two halves of a circle separated by a wavy line, there are two small dots of opposite colour in each half. These are said to represent the fact that there is a little bit of the opposite in everything. To me they have a radically different explanation. The black half of the wheel, for that is what it is, is a container filled with water upto the point where the small hole of opposite colour is. Excess water flows out through the little hole. The weight of the water makes the yin-yang symbol settle with the black water-filled half, downwards, but with the 'tadpole's tail slightly lifted. Now the white half is also filled with water to the point where excess water flows out through the little hole. Looking at the diagram it is clear that there is a tendency for the whole arrangement to tip over, thus causing a repeat action and leading to a gradual turning of the wheel. If there is insufficient impetus in this arrangement a very interesting addition can be made. If one considers the yin-yang symbol with its water inside as one part of the machine, and one then places another identical yin-yang on the same axis or axle, but fixed so that the central wavy line is precisely 90% further round, the same reaction will occur but more smoothly. And there is a bonus. Looked at from the side, but so that you can see through the first yin-yang symbol, to the adjoining one, one can see a swastika, but one with curved arms instead of straight ones! In addition to this the Chinese associate the yin-yang symbol with five elements -metal, wood, water, fire and earth. Metal for the bearings; wood for the construction; water for the weight; fire for smelting the metal, and earth to provide the ore, or to provide an alternative weight.

If one draws the yin-yang symbol with its two semi-circles too exactly semi-circular it becomes clear that the above description of the potential self-turning wheel does not work. The wavy line must be very gradual to enable the water to flow readily from one side to the other. This is of course intensly speculative but interesting nevertheless. It shows that it might be possible for an Orffyrean machine to have existed in remote ages. The similarity of the double yin-yang symbol to the swastika and the similarity of the single yin-yang symbol to the savonius windmill and to Orffyreus' windmill is also highly significant.

It is a curious fact that when I first came across the legend of Orffyreus' wheel, there were several accounts of the story available to me, mostly based on the original Strieder version; now, in the many compilations of 'Scientific

Mysteries'; 'Unsolved Mysteries'; Mysteries of the Past' etc., nowhere is there to be found any reference to Orffyreus' wonderful machine. Why this should be, I cannot tell. Can it be that the theory of perpetual motion has been so thoroughly denounced that no one dare suggest that it is even at least as likely as crop circles, UFO's, regression, telekinesis, thought reading, psychometry, ghosts, ET's, homeopathy, psychic healing, ley lines, astrology, aliens . . .? I dare so to do, and I suggest it has a considerably more logical and practical basis than those other three-dimensional mirages pushed at us in books, magazines, television and films.

Let research begin then, and a way be found to duplicate Orffyreus' machine. The benefits are of such staggering potential that it would be a crime against humanity to delay any longer. Let all those who believe perpetual motion (in the Offryrean sense) is impossible suspend their disbelief for a while. Let the dreamers and the mad professors and all those who labour in secret in their sheds and outhouses experimenting with new ideas, have their say and maybe a way will be found. And if after every avenue of research has been exhausted and no stone has been left unturned, no method of operating a wheel through the force of gravity has been found, then the search should be started again. Because I believe that such a device is possible. Johann Bessler discovered it nearly three hundred years ago and no one listened — listen now!

THE END

Appendix A

The first encoded passage.

Those who are keen to ask questionsshould ask them of this little book. My deeds will not be revealed prematurely. Should anyone wish to speculate about the truth, let him ponder on the rich pageantry of words which I now cause to shower down upon him!

For greed is an evil plant. An anvil receives many blows. A driver drives. A runner runs. The seer sees. The buyer buys. The rain drips down. Snow falls. The shotgun shoots. The bow twangs. A great herd of fat, lazy, plump horses wanders aimlessly. The flail would rather be with the thresher than the scholar. Children play with heavy clubsamong the broken columns. Acrobats and shadow-boxersare as fleet and nimble as the wind. The cunning cat slinks silentlyalong and snatches nice juicy mice. The dog creeps out of his kennel just as far as his chain will stretch. He knows how to please by playing with his little toys and knick-knacks. He wags his tail, creeps through the hoopand is rewarded with a pat on his paws by the stiff fops who watch him. A wheel appears — is it really a wheel, for it does not have a normal rim. It revolves, but without other wheels inside or outside, and without weights, wind or springs. Seen sideways or full-face it is as glorious as a peacock's tail. It turns to the right and to the left; it spins around in any direction, whether laden or empty. All things belong to one of the three kingdoms (animal, vegetable or matter) and you have the physical evidence before you. Without such things as sulphur, salt, and mercury all things will come to a standstill — the qualities of the elements are necessary to keep things going. Saturn, Mars and Jupiter are ready to join in any battle. Even the things we eat do not lose their elemental influence — for it spreads itself through every limb and sinue of our bodies. A crab crawls from side to side. It is sound for it is designed thus. Poltergeists wander freely through locked doors. But softly! — speak softly of all these marvels, lest the enemy grows wise! He will drench me with his spittle so that I will lose my temper and in a sudden fit, cast aside the mantle that conceals my wheel. But he shall be thwarted in his desires.

Der Geitz ist eine Wurzel Kosz
Eine Umbosz crieget viele Stoszt
Ein fuhrman fahrt/ ein Lausser lausset/
Eine Seher sicht/ der kausser tausset/
Der Regen fliess't/ der Schnee der fallt/
Die Buchse schiess's/der Bogen schnelle;
hier wandern eine grosse heerde
Sehr sette/ faute/ dicte pferde/
Die flegel wollen insgemein
Ben Dreschern/ nicht bey Doctern seyn;
Die kinder spielen aus den Saulgen
Mit lauter schweren Schniebe=Kaulgen;
Lusst=Springer Feder Fechter sind
Beschwind und hurtig wie der Wind/
Die schlaue Katze schleicher leise/
Und hascher seine sette Mause/
Der hund auch ausz der hutten creuch't/
Doch nur so weit die kette reich't/
Die schonen Schdagen und Machinen
Weisz er sehr freundlich zu bedienen/
Er wadelt wohl mit seinem Schwiess/
Kriecht aus dein Bauche durch den Reiss/
Dasur ihr bald die durren Poppen
Auch ziemlich aus die pfoten floppen;x
Man sicht ein Rad/ und auch sein Rad/
Weild Felgen und auch sein hat/
Lausst ohne in'und auszre Rader/
Zimbel=Bewicht/ Wind und Uhr=Feder/x
hier sichtets halb/ dort sichtets gantz/
Es prahlet wie ein Psauen=Schwantz/ x
Es lauss't zur Rechten und zur Lincten/
Man darff ihm nur mit Fingern wincten/
Es breitet sich die Lang'und Quer'/
hier ist es voll/-dort ist es leer';
Ein Ding besteh't ausz den dren Reichen/
Ihr habet: hand=greiffliche Zeichen/

Ohn' Schwesel/ Salz/ Mercurius
Auch bald ein Ding verfliessen wusz. x
Der Elementen Qualitreten
Auch iederm Dinge sind vonndhten;
Saturnus/ Mars/ Jupiter sein
Zu iederm Krieze willig seun. x
Das Ding auch/ (wovon man sich nahret)
Durch Darm und Marct und Beine fahret; x
Ein Krebs vormerts und ructlings triecht /
Und ist gesund (wohl zugericht.)
Die Polter+Geister fren spasziren
Zum offtern durch verschloss'n Thuren;x
Zwar stille/ stille/ und genug/
Genug/ sont wird der Feind nur flug/
Der Feind/ als der mich wil begensern/
Begensern/ ich sol mich erensern/
Erensern und aus frischer That
Den Mantel werssen von dem Rad; x
Doch dieser Wunsch wizd ihm miszlingen/

The above rendering of the german text may contain some inaccuracies due to the difficulty in reading the aged printing. The ink has smudged in some places, and made the deciphering of the actual text ambiguous, but in each instance it may only be one or two letters which are in doubt. In addition the limitations of my pc prevent me from reproducing the exact two/three letter combinations commonly used in german text and so I have reproduced them as near as is possible. The addition of the apparently random x at the ends of some lines is another feature reproduced as in the original, but inexplicable to me at this time.

In the page prior to this one he brings to our attention again, the date of 1717 — how relevant that is remains to be seen. But it seems evident that it is a pointer to the use of one of the hebrew codes discussed in chapter eleven. My feeling is that the important part of the above passage begins with the words, "For greed is an evil plant..." and possibly ends at "...the stiff fops who watch him".

Appendix B

This part contains all the Biblical references contained in the section entitled Orffyreus' declaration of faith. I have added any comments or finds concerning any of the text which might be helpful to those who wish to attempt to decode the cipher.

Page 122	**Page 123**	**Page124**	**Page 125**
Rom.1.v.16	Joh.v.16	Heb.10.v.25	1.Cor.10.v.31
Math.17.v.5	2.Cor.11.v.1	Psalm.68.v.27	Apoc.3.v.20
Matth.5.v.3-11	Psal.37.v.37	Matth.6.v.33	Matth.25.v.34
Act.4.v.12	Marc.12.v.30	2.Co.6.v.17	Tim.1.v.5
1.Cor.1.v.12	Sap.2.v.20	Act.16.v.31	Phil.3.v.2
Tit.2.v.12	Psal.11.v.2	1.Thess.5.v.21.24	Jac3.v.15
Eph.4.v.13	Matth.7.v.7.8	2.Tim.2.v.8	Col.3.v.17
Rom.14.v.18	Matth.5.v.44	Sap.6.v.19.20	2.Petr.1.v.5.6.7
Micha.6.v.8	Micha.7.v.9	1.Pet.4.v.5	Deut.12.v.52
Phil.4.v.13	Matth.5.v.16	2.Thess.2.v.3.4	Gal.5.&.16.
1.Cor.3.v.11	Psal.4.v.4	Psalm.31.v.7	Tit.2.v.7.8
Col.2.v.11	Jac.1.v.17	1.Pet.1.v.19	Eph.3.v.19
Esa.53.v.5	1.Tim.3.v.16	Col.2.v.3.4	Ephes.2.v.9
Ezech.16.v.9-14	1.Joh.2.v.15.16.17	2.Tim.3.v.12	Rom.3.v.24.28
Tit.2.v.14	Psal.119.v.172	Gal.1.v 20	Apoc.3.v.16
Joh.14.v.6	Rom.10.v.9.10	2.Reg.17.v.13	1.Cor.10.v.12
2.Cor.13.v5	Jer.6.v.16	Hos.6.v.1	Jac.1.v.26
Gen.8.v.21	Joh.17.v.23	Tit.2.v.11	Jac.2.v.17.18
Job.9.v.3	1.Cor.10.v.31	1.Joh.3.v.1	Hebr.3.v.12
	Jac.2.v.13	1.Petr.5.v.8.9	Hebr.10.v.24
	1.Pet.2.v.17	1.Tim.6.v.12	Hebr.4.v.1
	Rom.13.v.1.7	Prov.28.v.13	Matth.16.v.26
	1.Pet.1.2.v.17	1.Cor.11.v.26	1.Petr.4.v.3
	Matth.22.v.21	Matth.16.v.24	Hiob.7.v.1
	Levit.19.v.32	Luc.21.v.24	
	Heb.13.v.17	1.Tim.3.v.16	
		Joh.17.v.3	

Page 126	Page 127	Page 128
1.Rom.8.v.33	2.Tim.3.v.16	Esa.6.v3
Psal.116.v.13.14.	Rom.10.v.4	2.Tim.4.v.18
1.Cor.1.v.30	Joh.1.v.1.14	Matth.25.v.41.
Eph.4.v.24	B.Weisch.7	1.Petr.v.9
Eph.1.v.5.6.	Rom.8.v.9.14.15.16.26	
Matth.5.v.48	Joh.5.v.7	
Psal.5.v.5.6.7	Rom.3.v.24.25	
Prov.23.v.20.21	Rom.14.v.7.8	
Eph.4.v.2	Thren.3.v41	
Apoc.3.v.11	Phil.3.v.20	
Psal.12.v.4.5.	Jer.13.v.23	
Matth.7.v.13	Prov.3.v.5.	
Jac.3.v.5.11	Psal.37.v.3.4.5	
Phil.2.v.12	Psal.95.& 103	
Matth.9.v.2	2.Cor.9.v.15	
Jer.31.v.8.	Psal.118.v.17.18	
Tit.3.v.5.6.7.		
Gal.3.v.26		
Joh.3.v.5		
Gal.5.v.19.20.21.		
1.Joh.3.v.9		
Tit.2.v.11		
1.Petr.2.v.9		
1.Cor.3.v.18		

Each column has been copied as it appears in Bessler's book, one column to a page. The variations in the spelling of the Biblical books quoted in the references, is as per the original document. I have not included the German comments which appear along side each reference, but those who wish to have copies in order to assist in deciphering the code can apply for one either from me or from the British Library.

The first problem lies in identifying which Bible Bessler used. My first thought was that it must be the Lutheran Bible, but then when I compared the text I discovered variations between the Luther Bible and the text actually quoted by Bessler where he has done so. Also some of the Biblical references only appear in the Apocrypha. Some of the Book Titles are confusing — Jac. refers to James for instance, and Hiob refers to Job. Other problems involve for instance Psalm 11 in the Lutheran Bible is Psalm 12 in the Vulgate version; Psalm 37 in both the Lutheran and King James version has 40 verses, but only 23 in the Vulgate version. In summary then Bessler did not use the Vulgate version because he quotes references which do not appear, but on the other hand he did not use the Lutheran Bible because he quotes from the Apocrypha which was not included in that Bible and the text differs anyway where it is matched.

There may be a clue to which Bible he used because just prior to the seven pages of Biblical references he misquotes deliberately, a verse from the Bible. He says "I am neither of Paul, nor of Apollo, nor a Papist, nor of Cephas, nor of Luther! What then? I am of Christ." I take this to mean that he does not use the Pauline Bible, nor the Greek, Roman Catholic, Orthodox or Lutheran versions. I think that there was a local version available to him but which is difficult to say. There were many Bibles in circulation at this time and it might be possible to identify the exact one by matching the quoted text to a particular one.

Suprisingly Bessler quotes the same Biblical reference twice in five places. This may indicate as I said earlier that some word, difficult to find more than once, was required. The relevant references are as follows;

Tit.2.v.14
1.Tim.3.v.16
1.Cor.10.v.31
Tit.2.v.11 and
Rom.3.v.24

If photocopies are required of the original coded passages it is only fair to warn anyone considering purchasing these that some of the text is difficult to discern due to the age of the original and the inadequacies of photocopying. If you still require copies please feel free to contact me either by mail at ;

Permo Publications
PO Box 2001
Leamington Spa
CV32 6YQ
United Kingdom (UK)

or email me at
jcollins@free-energy.co.uk

If you require a further copy/copies of the book please send email to
permo-pubs@free-energy.co.uk

Bibliography

A Course of Experimental Philosophy, vols I & II by J.T. Desaguliers. London 1734
Acta Eruditorum 1681 - 1731,1732-1776, Leipzig, British Library
An Account of the Automaton constructed by Orffyreus in two letters.....ed. by Dr. W. Kenrick, London,1770
An Account of the Courts of Prussia and Hanover by John Toland 1705
"The case of the Learned represented according to the merit of the ill progress hither to made in Arts and Sciences; shewing i)-The cause of gravity and attraction, ii)- What Nature is and the effects it is capable of" by John Conrad Francis de Hatzfeld,1724. British Library
Annal.Physico-Med. de Breslau, Leipzig 1723
An Outline of the career of J.T.Desaguliers, Wilfred Hurst 1928
"Astronomical Dialogues between a gentleman and a lady: wherein the doctrine of the sphere, uses of the globes, and the elements of astronomy are explained in a pleasant, easy and familiar way. With descripotion of the famous instrument called an Orrery" - John Harris. 2nd edition, London 1725
At the Sign of the Orrery, the Origin of the Firm of Cooke, Troughton & Simms Ltd, by E.W. Taylor & J.S. Wilson 2nd Edition, York, 1969
Beyond 2001- The Laws of Physics Revolutionised. by Sandy Kidd, Sidgewick and Jackson 1990
Biographical Memoires of the Royal Society, vols 1-33
Briefe von Christian Wolff aus der Jahren 1716-1753, St. Petersburg 1860
Leibnizens und Huygens' Briefweschsel mit Papin. Berlin, 1881 by Ernst Garland
Calender of Treasury Books and Papers
Century of Inventions by Edward Somerset (Marquis of Worcester), London, 1655
Charles Bridgeman and the English Landscape Garden by Peter Willis, A. Zwemmer
Clockmaking in Oxfordshire, 3rd edition, 1989
Das Perpetuum Mobile by A. Daul, Vienna, 1900
Das Triumphirende Perpetuum Mobile - 1719 - Orffyreus, pseud(i.e. J.E.E. Bessler). British Library.
Der Rechtglaubige Orffyreer: oder die... Orffyreus. pseud (i.e. J.E.E.Bessler.)
Diary of Sir Dudley Ryder

Diary of Humfrey Wanley, 1722
Diary of Ralph Thoresby. FRS 1677-1724 by Rev. Joseph Hunter FSA in 2 vols London
1740
Diary of Mary, Countess Cowper 1714 - 1720, ed. S.Cowper (2nd edit.) 1865
Encyclopaedia Brittanica
Fischer von Erlach der Jungere by J. Schmidt
Foibles and fallacies of Science by Daniel Herring, Van Nostrand, New York, 1924
Frederick the Great, by Alan Palmer,George Weidenfeld & Nicholson, London 1974
Gardner to Queen Anne, Henry Wise (1653-1738) by David Green
George I, by Ragnhild M. Hatton, Thames and Hudson 1978
Grundlage zu einer hessischen Gelehrtenund Schrifstellergeschichte by Friederich W. Strieder
Gruendlicher Bericht, von dem Durch Den.....Mathematician Herr Orffyreanum glucklich inventiren Perpetuo ac per se Mobile, nebst accurater Abbildung - 1715 - Johann Theod. Boetio. British Library.
Historical measures, weights, calendars and moneys of all nations by W.S.B. Woolhouse, London 1890
Historical Register 1718,1719, 1720
Historie de Gelahrteit unserer by Frederich Wilhelm Strieder
Hofdiarum of Hertzog Moritz-Wilhelm von Sachsen-Zeitz at Sachsisches Staatsarchiv, Dresden
J.E.Fischer von Erlach by T Zacharias Vienna and Munich 1960
Landgraf Karl von Hessen-Kassell by Hans Philippi 1976 Marburg
Leibniz: An Introduction to His Philosophy (1979) by Nicholas Rescher
Leibnizin seinen Bezichungen zu Russland und Peter the Great, by Guerrier
Letters Historiques. Amsterdam, 1721
Letters of Samuel Johnson, ed. Dr.Birbeck Hill II
Life of J.T. Desaguliers by J.D. Stokes 1952 vol 38
Life of Steele, G.A. Aitken 2 vols 1889
London in 1710, from the travels of Zacharias Conrad von Uffenbach, translated and edited by W. M. Quarrel and Margaret Hare, Faber & Faber,London, 1934
Mathematisches Lexicon by Christian Wolff, Leipzig.1716
Mathematical practitioners of Hanoverian England by E.G.R.Taylor. Cambridge1966
Mathematical Magick by Bishop John Wilkins, London 1648
Memoires of the House and Dominions of Hesse-Cassel: giving an account of the origins and history of that family with political remarks, London, 1740
Nauka i literature v Rossii pri Petre Velikom (Science and Literature in Russia under Peter the Great), by P. Pekarskii. 2 vols St. Petersburg, 1862 (vol I Pages 33 &34, 49-52, 533-558)
Neue Nachricht von der curiosen und... - 1718
Newton's sketches of perpetual motion - Add MS 3996,103, Add MS 3996,ff 121, 102 British Library
Oddities- A Book Of Unexplained Facts, by Rupert T. Gould, RN, Philip Allan &

Co, London, *Cornelius Drebbel: A Neglected Genius of Seventeenth Century Technology, Presidential Address by L. E. Harris delivered at the Science Museum, London, 8th October 1958 1928*
Oevres Philosophiques et Mathematiques de M.G.J. 's Gravesande, rassemblees et publiees par J. N. S. Allamand. Amsterdam, 1774,
Orffyreus Apologische Poesi und Poetische Apologie von seinem mobile per se un perpetuum mobile - 1716/17 - Orffyreus, pseud (i.e. J.E.E.Bessler). British Library.
Perpetual Motion -The History of an Obsession by Arthur W.G. Ord-Hume. Allen & Unwin. 1977
Perpetuum Mobile, by Henry C.E. Dirks, E & F. Spon, London, vols I & II 1861/ 1870
Peter the Great by Ian Grey, 1960, New York
Peter the Great by V. Klyuchersky, London, 1958
Political State of Great Britain vol XVII 1720
Science and Literature in Russia in the Reign of Peter the Great, by P. Pekarsky, St. Petersburg 1862
South Sea Bubble J. Caswell
*Theatrum Machinarum Generale.*by Jacob Leupold, Leipzig,1724
The 'Englishman' -ed Richard Steele -Thurs, October 29, 1713
The History of the King's Works vol V by H.M. Colvin
The Memoirs of the Marquess de Langallerie by Gatien de Courtilz de Sandras
The Perpetual Motion Mystery, R. A. Ford (Lindsay Pub) 1986
The Seven Follies of Science by John Phin, Constable, London 1913
The Steam engine of Thomas Newcomen by L.T.C.Rolt & J.S.Allen, New York. 1997
Wolff's Mono Writings, Halle 1755
"Virtuosi or St. Luke, a Clubb in London Club" - folio tract. Add.MS.39167B folios 73-86 at the British Museum

Useful addresses

Landesmuseum Staatliche
Kunstsammlungen
Bruder-Grimm-Platz 5
D - 3500 Kassel

Handschriften der Alten
Gesamthochschulebibliothek Kasel
Landesbibliothek U
Murhardsche Bibliothek der
Staatkassel
Monchbergstasse 19
3500 Kassel, Hessen

Stadtarchiv Kassel
Rathaus
Postfach 3500
Kassel
Germany

Niedersachsisches Hauptstaatsarchiv
Stadtarchiv Kassel
Rathaus
Postfach 3500
Kassl
Germany

Am Archiv
30169 Hannover
Germany

Geheimes Staatsarchiv Preusischer
Kulturbesitz
Archivstr. 12-14
14195 Berlin
Germany

Nieders. Staatsarchiv
Forstweg 2
38302 Wolfenbuttel
Germany

Hessisches Staatsarchiv
Friedrichsplatz 15
W - 3550 Marburg
Germany

Braunschweigisches Landesmuseum
Burgplatz 1
D-3300 Braunschweig
Germany

Verkehrsamt
Haus des Gastes
3476 Furstenberg
Germany

The Royal Society
6 Carlton House Terrace
London SW1Y 5AG

Index